Athletics and Mathematics in Archaic Corinth:

The Origins of the Greek *Stadion*

Athletics and Mathematics in Archaic Corinth:
The Origins of the Greek *Stadion*

David Gilman Romano

iv

Memoirs
of the
American Philosophical Society
Held at Philadelphia
For Promoting Useful Knowledge
Volume 206

Library of Congress Catalog Card No.: 92-75705
International Standard Book No.: 0-87169-206-6
US ISSN: 0065-9738

TABLE OF CONTENTS

vi

Dedication

To My Students at the University of Pennsylvania —

Past, Present and Future

LIST OF ABBREVIATIONS

These abbreviations generally follow those suggested by the *American Journal of Archaeology* Vol. 95, 1991, pp. 1-16.

ABV = J.D. Beazley, *Attic Black-Figure Vase-Painters*, Oxford, 1965.

AJA = *American Journal of Archaeology*.

AM = *Mitteilungen des Deutschen Archäologischen Instituts, Athenische Abteilung*.

ArchDelt = Ἀρχαιολογικὸν Δελτίον.

ASAtene = *Annuario della R. Scuola Archeologica di Atene*.

BCH = *Bulletin de correspondence hellénique*.

Broneer, *Isthmia II*, 1973 = Oscar Broneer, *Isthmia*, Volume II, *Topography and Architecture*, Princeton, 1973.

Broneer, *South Stoa*, 1954 = Oscar Broneer, *Corinth*, Volume I, Part iv, *The South Stoa and Its Roman Successors*, Princeton, 1954.

FdD = *Fouilles de Delphes*, École Française d'Athènes.

Heath, 1921 = Sir Thomas Heath, *A History of Mathematics*, Volume I, From Thales to Euclid, Oxford, 1921 (republished New York, 1981).

Hesperia = *Hesperia,* Journal of the American School of Classical Studies at Athens.

Humphrey, 1986 = John Humphrey, *Roman Circuses, Arenas for Chariot Racing,* Berkeley, 1986.

IG = *Inscriptiones Graecae.*

JdI = *Jahrbuch des k. deutschen archäologischen Instituts.*

Mind and Body = Olga Tzachou-Alexandri, ed., *Mind and Body, Athletic Contests in Ancient Greece,* Athens, 1989.

Neugebauer, 1969 = Otto Neugebauer, *The Exact Sciences in Antiquity*, Second Edition, New York, 1969.

ÖJh = *Jahreshefte des österreichischen archäologischen Instituts.*

OlBer = *Bericht über die Ausgrabungen in Olympia.*

OpusAth = *Opuscula Atheniensia.*

Praktika = Πρακτικὰ τῆς ἐν Ἀθήναις Ἀρχαιολογικῆς Ἑταιρείας.

Raschke, 1988 = Wendy J. Raschke, ed., *The Archaeology of The Olympics*, Madison, Wisconsin, 1988.

RE = Pauly-Wissowa, *Real-Encyclopädie der klassischen Altertumswissenschaft.*

Romano, "Stadia," 1981 = David Gilman Romano, "The Stadia of the Peloponnesos," dissertation in Classical Archaeology, University of Pennsylvania, 1981, University Microfilms, Ann Arbor, Michigan.

Romano, "Arete," 1983 = David Gilman Romano, "The Ancient Stadium: Athletes and Arete," *The Ancient World* 7 (1983): 9-16.

Williams, 1980 = Charles K. Williams, II and Pamela Russell, "Corinth Excavations of 1980," *Hesperia* 50 (1981): 1-44.

Romano

PREFACE

This work is a study of the origins of the ancient Greek stadium, especially with regard to the archaeological evidence from the Archaic and Classical sites of Corinth, Isthmia, Halieis and Olympia. The earliest remains of the Greek *stadion* come from the Peloponnesos, a region of southern Greece, although the architectural structure eventually became well known all over the Greek and Roman world. I also include the ancient evidence for the initial appearance of the word *stadion* in the Greek language and its early use in the sixth and fifth centuries B.C. Aspects of this study formed the basis of my doctoral dissertation, "The Stadia of the Peloponnesos," submitted to the University of Pennsylvania in 1981, an architectural study of the origin, development and design of the Peloponnesian stadia from the sixth century B.C. through the fourth century A.D.

The primary component of the current work is the most recent archaeological research from Ancient Corinth concerning the Archaic *dromos* and the Early Classical starting line and its significance for the study of Greek and Roman athletics, as well as the understanding of early Greek mathematics. The field work for this study was carried out as a part of the Corinth Computer Project, begun in 1988, which is supported by the Corinth Computer Project Fund of The University Museum as well as by the School of Arts and Sciences, the Department of Classical Studies and the Graduate Group in Classical Archaeology of the University of Pennsylvania. The project has also received generous support from Autodesk, Inc., especially from Jim Purcell and his colleagues in the Education Department, the IBM Corporation, the Lietz Corporation (Sokkia) and Softdesk (DCA Engineering). I thank the Director of The University Museum, Robert H. Dyson, for his interest in this project and Gerald J. Porter of the School of Arts and Sciences and the Penn - IBM Threshold Project for his assistance and support over the past several years. Although the Corinth Computer Project has had as its principal objective the survey and study of the colonial city plan of Roman Corinth, numerous related issues have been raised resulting in studies such as this one.

I am especially grateful to Charles K. Williams, II, Director of the Corinth Excavations, American School of Classical Studies at Athens, for permission to study and publish this report on the Corinth material. It was he who first introduced me to the subject of the Corinth racecourses as a student at the American School of Classical Studies in 1976. I also thank him, as well as the Assistant Director of the Corinth Excavations, Nancy Bookidis, for kind assistance in all aspects of this study including permission to reproduce a number of drawings and photographs.

I am also indebted to the late Oscar Broneer for many discussions on the subject of the ancient *stadion* in general, and of Isthmian and Corinthian questions in particular, and to him and to Elizabeth R. Gebhard for permission to reproduce photographs and drawings from the University of Chicago Isthmia Excavations. I thank Timothy Gregory for permission to reproduce the modern drawing of the Sanctuary of Poseidon from the most recent excavation and survey by The Ohio State University. I thank the late Alfred Mallwitz for discussing with me in 1984, at some length, the successive stadia at Olympia and for permission to reproduce photographs and plans

1. Corinth Computer Project team, 1989. From left to right, Christopher Campbell, Mimi Woods, the author, Elizabeth Johnston and Benjamin Schoenbrun. A portion of the Early Classical starting line is visible in the foreground. Photo courtesy of Irene Bald Romano.

from Olympia. I also thank Michael H. Jameson for permission to include the material from Halieis in this study. For other reproduction rights I thank Carlos A. Picón of the Metropolitan Museum of Art, the Deutsches Archäologisches Institut, Athens, the Deutsches Archäologisches Institut, Rome, the American School of Classical Studies at Athens, B.F. Cook of The British Museum, Olga Tzachou-Alexandri of the National Archaeological Museum, Athens, the Archaeological Society of Athens and the Johns Hopkins University Press. For photographic assistance I thank H. Fred Schoch of The University Museum.

During the past five years at the University of Pennsylvania I have been aided by two student research assistants who have devoted some of their time to the further understanding of the successive Corinth starting lines: David Conwell, graduate student in Classical Archaeology, during the year 1986-1987, and Benjamin Schoenbrun, Corinth Computer Project Research Intern at The University Museum for the years 1989-1991. The latter is responsible for the execution of all of the computer generated drawings in this study, unless otherwise noted. During the summers of 1988 and 1989 I was assisted in the field with measurements and other associated work by members of the Corinth Computer Project team, under my direction: Douglas Arbittier, Christopher Campbell, Michael Foundethakis, Elizabeth Johnston, Benjamin Schoenbrun and Mimi Woods, all of the University of Pennsylvania (Fig. 1). I thank them for their hard work, good humor and constructive contributions toward the success of this project. I have also discussed aspects of the design of the Greek starting lines in Corinth with other scholars including Stephen Glass, A. John Graham, John Humphrey, Michael H. Jameson, John Keenan, Antony Raubitschek, Robin Rhodes, Homer A. Thompson, Charles K. Williams and Thomas

H. Wood. I thank them for their time and interest. For assistance with the history of Greek mathematics, I thank A. John Graham and Wesley Smith. I am grateful to Åke Sjöberg and Erle Leichty for questions concerning Mesopotamian influences and David O'Connor and Janet Richards for assistance in Egyptian matters.

It is important to note at the outset that much of the field and laboratory research for this study, specifically the work on the Corinth material, could not have been successfully accomplished without the assistance of the computer and the commercially available architectural, engineering and surveying programs (see Chapter 2, page 53, note 26) which have been utilized. Specifically, the ability to be able to carry out an architectural field survey to a very high degree of precision has meant that the accuracy and reliability of the field data was never in doubt. Furthermore, the ease with which the architectural and engineering computer programs can be put to use in the study of the architecture and monuments of the ancient world has brought a whole new understanding to the field, and this work is a product of the kind of sophisticated measurement and analysis now available to the modern archaeologist and scholar. The color drawings that accompany this work have been produced by the computer using AutoCAD, and I thank Autodesk, Inc. for a generous subvention toward the publication.

I am especially grateful to my family, to my wife Irene, our three daughters, Katy, Elizabeth and Sarah, and to my father and my late mother for their support and encouragement during the many hours devoted to this project over the past several years and the numerous summers spent in Greece. Finally, I am indebted to the American Philosophical Society for the opportunity to present the results of the Corinth research at its Autumn General Meeting, November 9, 1990, and for its willingness to publish this study.

September 15, 1992
Philadelphia

INTRODUCTION

How common the word "stadium" is in today's world and how much it is a part of our daily lives! Hardly a day passes when we don't see an athletic event on television — baseball, football, soccer, track, tennis — from a stadium somewhere in the world attended by thousands and watched by millions. Modern stadia have grown to tremendous proportions of size and scale and now can often include seating accommodations for between 50,000 - 100,000, artificial grass, all weather running surfaces, computerized scoreboards and even air conditioning and climate control in an indoor setting. Multi-purpose facilities have become common in many modern cities where there is a single building that can be used for different events at all times of the year including basketball, baseball, football, hockey, track and field, tennis, field lacrosse, as well as meetings and expositions of different kinds. Many modern facilities also include luxurious "box seats" or suites, sometimes high above the playing field, replete with lounging and dining areas, bars and other modern amenities.

Among the largest and most prominent buildings of our modern cities are the public stadia and athletic facilities that are oftentimes linked with the pride and importance of the city itself. In the 1960s a generation of new stadia for professional athletic teams were built in some of this nation's largest cities, including for example, St. Louis, Missouri, where the new Busch Memorial Stadium was a key element of downtown urban renewal. A similar phenomenon may be happening again in the 1990's, for example the new Oriole Park at Camden Yards in a rejuvenated Baltimore Harbor district. In addition, the domed athletic facilities of the past almost thirty years have brought a great deal of prestige to the cities where they are located.

Beginning with the Astrodome in Houston in 1965, these facilities are now fairly commonplace in the modern world and even have an ancient predecessor in the Roman amphitheater, in terms of overall shape and design, and spectator capacity. The awning of the ancient structure, the *velarium*, suggests the enclosed nature of the modern building.[1]

Equally important in the modern day although perhaps less recognized and certainly less celebrated, are the thousands of smaller college, high school and elementary school stadia which, while modest in design and scale, are closer to the original concept of the early Greek *stadion*: a simple track or playing field bordered on one side by a grandstand or at times an embankment of earth, and designed principally for the use of the competitors.

What is the origin of the stadium, what was its original design and function and how did it begin in ancient Greece? Although one may think of the ancient stadium as a secular facility, similar to its modern counterpart, the ancient stadium was originally a religious structure. It was commonly found in both urban and rural settings, as a part of Greek sanctuaries, and used in connection with cult practices; it was used to accommodate athletes and spectators for athletic contests.

The *stadion* was used specifically for human athletic contests whereas the Greek *hippodrome* and later the Roman circus were used for equestrian events. The *gymnasion* and the *palaistra* were used for training purposes for human athletic events. Contests commonly held within the *stadion* included footraces, e.g., the *stadion* (one length), the *diaulos* (two lengths), the *hippios dromos* (four

---

1. For the Roman amphitheater see Katherine Welch, "Roman amphitheatres revived," *Journal of Roman Archaeology*, 4 (1991): 272-281 for a review of two recent publications on the Roman amphitheater; Jean-Claude Golvin, *L'Amphithéater Romain. Essai sur la Théorisation de sa Forme et des Fonctions*, Paris, 1988, and *Spectacula-I. Gladiateurs et Amphithéatres. Actes du Colloque à tenu " Lattes les 26, 27, 28 et 29 Mai 1987*. Lattes. See also Augusta Hönle and Anton Henze, *Römische Amphitheater und Stadien, Gladiatorenkämpfe und Circusspiele*, Zurich, 1981.

2. Olympia, bronze statuette, B 26, ca. 490 B.C. Photo DAI, Athens, neg. nr. Olympia 811.

lengths), the *hoplitodromos* (a race in armor of two or four lengths), the *dolichos* (a distance race of multiple lengths) as well as field events such as boxing, wrestling, *pankration,* which was a combination of wrestling and boxing, or the five events of the *pentathlon: diskos,* javelin, long jump, wrestling and the *stadion.* These athletic contests were religious in the sense that each athlete, through the discipline of physical training and competition, made an individual offering to the god or hero in whose honor the games were held. The achievement of excellence or *arete* in the form of a victory in one of the contests was the ultimate expression of honor to the god. This may be best symbolized in the form of a miniature bronze statuette as an athlete's dedication from the Sanctuary of Zeus at Olympia, dating to ca. 490 B.C., depicting an athlete at the start of a footrace. On the right leg of the figure is inscribed the words, τὸ δίϝος ἰμί, "I belong to Zeus" (Fig. 2).[2]

It is, therefore, no coincidence that elements of the earliest stadia from the Greek world have been found in the Archaic Greek Panhellenic sanctuaries of Olympia and Isthmia, situated close to the major temple and the principal altar at each site and dating to the mid-sixth century B.C. (Fig. 3). Although very little is left of these early stadia - only the partial remains of the retaining walls that once supported artificial earth embankments which were constructed to provide spectators a place to sit or stand - there is enough to recognize the essential elements of the structures.

In addition to the earliest of archaeological remains of Greek stadia, there is from Corinth an example of an Archaic *dromos* with

2. Alfred Mallwitz and Hans-Volkmar Herrmann, *Die Funde aus Olympia,* Athens (1980): 156-157, no. 107.1. The inscription is written in retrograde. The height of the bronze including the base is 10.2 cm. This miniature bronze, Olympia, B 26, is associated with a similar piece from Olympia, B 6767, that depicts a discus thrower (Mallwitz and Herrmann, *op. cit.*, no. 107.2). There is a similar inscription. Both are believed to be of an Argive workshop. See also *Mind and Body,* no. 112, 113; pp. 221-222.

parts of the racecourse floor and an Early Classical starting line in situ.[3] As such, the Corinth *dromos* is the earliest datable starting line and preserved track surface from the Greek world. The nature of the design of the starting line indicates that it was built for the start of a distance footrace, *dolichos* or possibly a *hippios dromos* and much of the racecourse as well as the nature of the race itself can be reconstructed.

Better preserved are later, fifth century B.C. stadia, for example those at Olympia, Isthmia and Halieis which give us a good example of what a Classical stadium looked like. Each of these facilities had one and sometimes two artificial earth embankments that bordered the racecourse along its long sides. The ends of the *dromos* were limited by stone starting lines and the sides by a stone curb or water channel. These early stadia provided seats for only the judges and a few VIPs. The rest of the crowd either sat, or more probably stood, on the gradual earth slopes of the embankments. It may be from this use of the embankments as a place where spectators stood that the stadium took its name — "the standing place." These fifth century B.C. stadia were well within the limits of the sanctuary and were still located near the principal temple and altar of the site.

During the Late Classical and Early Hellenistic period, there was a trend to move stadia out of the sanctuary to a nearby location, probably in order to provide more extensive spectator space. These new stadia often took full advantage of the natural contours of a valley or hollow which provided a setting for spectators on the sloping hillsides. Examples from the Peloponnesos of such "second genera-

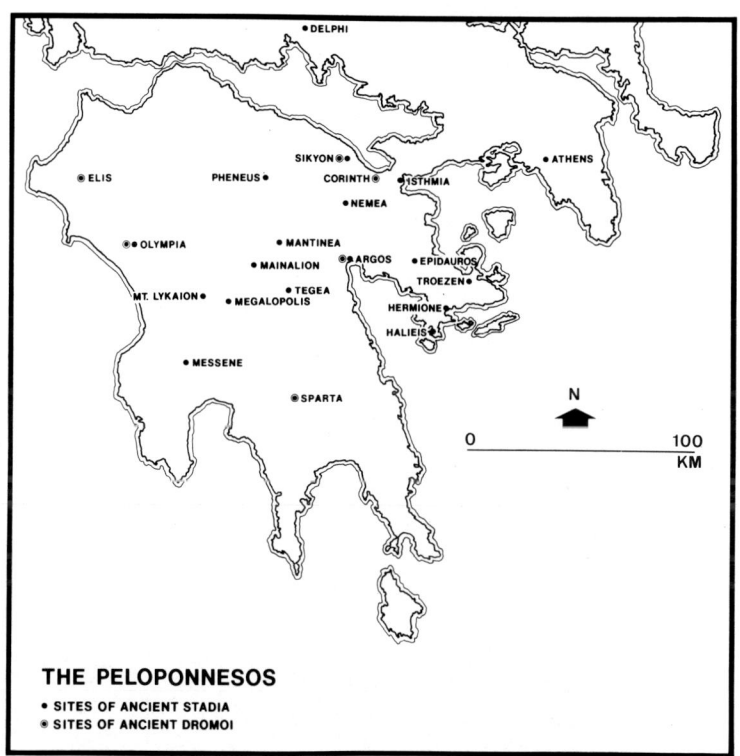

THE PELOPONNESOS
• SITES OF ANCIENT STADIA
⊚ SITES OF ANCIENT DROMOI

3. Map of the Peloponnesos, illustrating locations of ancient stadia and *dromoi* based on archaeological, historical, epigraphical and literary evidence. Drawing by Elizabeth Simpson.

3. A *dromos* is defined principally as a racecourse of a *stadion* in length with no formal facilities for spectators, see below p. 41.

4. Olympia III Stadium as seen from the northeast. The stadium has been excavated and reconstructed by the Deutsches Archäologisches Institut. Photo DAI, Athens, neg. nr. Olympia 5307.

tion" stadia are known at Olympia, (Fig.4) Nemea[4] and Isthmia. Instead of the rectilinear shape of the earlier *dromos*, the stadium floor was often characterized by two slightly convex long sides. A part of many of these new stadia of the Hellenistic period were vaulted entrances which provided direct access to the stadium floor from the nearby sanctuary. Such vaulted passageways are now known from the stadia of Epidauros,[5] Nemea and Olympia (Fig.5). Other Hellenistic and Roman stadia were combined with theaters in a variety of designs in order to increase spectators' accommodation and, as a result, probably created multi-purpose facilities. Examples from Asia Minor and East Greece are known from

4. The results of excavation of the late fourth century stadium at Nemea by the University of California at Berkeley, Stephen G. Miller, Director, have appeared in preliminary form in *Hesperia* from 1974 to the present time. The final publication series of the site is now in progress and the volume on the stadium is being prepared by Stephen G. Miller and David Gilman Romano. See also Romano, "Stadia" (1981): 71-114 and Stephen G. Miller, ed., *Nemea, A Guide to The Site and Museum*, Berkeley (1990), especially chapter 5: 171-191. Most recently see Stephen G. Miller, "The Stadium at Nemea and the Nemean Games," *Proceedings of an International Symposium on the Olympic Games (1988)*, William Coulson and Helmut Kyrieleis, eds., Athens (1992): 81-86.

5. For the stadium at Epidauros see P. Kavvadias, "Τὸ Στάδιον," in *Τὸ Ἱερὸν τοῦ Ἀσκληπιοῦ ἐν Ἐπιδαύρωι*, Athens (1900): 96-118, and P. Kavvadias, "Ἀνασκαφαὶ ἐν Ἐπιδαύρωι, Τὸ Στάδιον, *Praktika* (1902): 78-92 and Roberto Patrucco, *Lo stadio di Epidauro*, Florence, 1976.

5. Olympia Stadium vaulted entrance, as seen from the west. A portion of the vault has been reconstructed in the modern day. Photo DAI, Athens, neg. nr 72/3867.

6. The Panathenaic Stadium in Athens, originally built in the second century A.D. and reconstructed for the first celebration of the Modern Olympic Games in 1896. Photo by the author.

Aizanoi, Sardis, Tralles, Pergamon[6] and Rhodes as well as from Athens[7] and Dodona.[8]

In the Hellenistic and Roman periods seating facilities were more commonly, although not always, provided for spectators. Possibly the best known ancient stadium, that is still in use in the modern day, is the Panathenaic Stadium in Athens,[9] originally built by Herodes Atticus in the second century A.D. and restored and rebuilt largely on the Roman model in the late nineteenth century A.D. as the first site of the Modern Olympic Games in 1896 (Fig.6). In the modern day the spectator accommodation of this stadium is approximately 50,000. Herodes Atticus also renovated the then existing stadium in the Sanctuary of Apollo at Delphi, installing stone seats for the spectators.[10] The stadium at Olympia was renovated in the Roman period as well, although seats for the spectators were never installed.

6. David Gilman Romano, "The Stadium of Eumenes II at Pergamon," *AJA* 86 (1982): 586-589.

7. David Gilman Romano, "The Panathenaic Stadium and Theater of Lykourgos: A Re-examination of the Facilities on the Pnyx Hill," *AJA* 89 (1985): 441-454.

8. S.I. Dakaris, *Archaeological Guide to Dodona*, translated by Elli Kirk-Deftereou, Ioannina (1971): 72, fig. 23.

9. See Carlo Gasparri, "Lo stadio panatenaico," *ASAtene* 52-53, New Series, 36-37 (1974-1975): 313-392, and Jennifer Tobin, "The Monuments of Herodes Atticus," dissertation in Classical Archaeology, University of Pennsylvania, 1991. Most recently, see Jennifer Tobin, "Some New Thoughts on Herodes Atticus's Tomb, His Stadium of 143/4, and Philostratus VS 2.550," *AJA* 97 (1993): 81-89.

10. See Pierre Aupert, "Le Stade," *FdD*, Volume II, *Topographie et Architecture*, Paris 1979; Pierre Aupert, "Le cadre des jeux Pythiques," *Proceedings of an International Symposium on the Olympic Games (1988)*, William Coulson and Helmut Kyrieleis, eds., Athens (1992): 67-71.

The evidence that is presented below is arranged largely in chronological order. I first discuss possible precursors to the Greek stadium from the earlier civilizations of Mesopotamia and Egypt and then review the evidence for the earliest appearance and use of the Greek word *stadion* from literary, epigraphical and historical sources. What follows is a summary of the archaeological evidence for the earliest stadia from the Peloponnesos from Olympia, Isthmia and Halieis. The main body of the work is a discussion of the archaeological evidence from Corinth for the Archaic *dromos* and the Early Classical starting line, ca. 500 B.C., including a reconstruction of the racecourse, the nature of the footrace held there and a consideration of some of the athletes that may have used the *dromos*. This is followed by a brief discussion of our knowledge of Archaic Greek mathematics, and the implications for Greek geometry stemming from the archaeological evidence from Corinth. I then examine the mechanics of the starting lines from the fifth century B.C. stadium at nearby Isthmia, followed by a summary of the archaeological remains of the Hellenistic *dromos* and starting line in Corinth. Finally, I analyze the influence of the Early Classical Greek starting line in Corinth on the design of aspects of the Roman circus.

Chapter 1
ORIGINS AND EVOLUTION OF THE ANCIENT *STADION*

Ancient Precedents

This study is primarily concerned with the evidence for the beginnings of the *stadion* in ancient Greece. Although it is commonly thought that there was no precedent for the Greek structure from any earlier civilization, there is some literary, epigraphical and archaeological evidence to suggest that there existed in Egypt, and possibly in Mesopotamia, locations and facilities for athletic displays and contests.

Athletics are known from the third millennium B.C. in Mesopotamia where belt wrestling, a contest in which the athletes wrestled wearing only a belt, is well documented. The Old Babylonian version (ca. 1700 B.C.) of the Epic of Gilgamesh contains a well known wrestling scene where Gilgamesh and Enkidu fight.[1]

Also from the Old Babylonian period is the story of the marriage festival of Mardu, a western Semitic god, at which athletic contests are held in the courtyard of the temple.[2]

Shulgi, the second king of the famous dynasty of Ur (reign 2094-2047 B.C.) was known for many accomplishments, including athletics. Shulgi himself boasts of his athletic achievements which were won in the great courtyard of the temple.[3] He is also known as a great long distance runner and praises himself for his speed and endurance in running a round trip, in one day, between two religious festivals, one in Nippur and the other in Ur, some 100 miles distant.[4]

A Sumerian hymn was composed in honor of this feat and the king himself was said to have recited it. A portion of it follows:

> *That my name be established unto distant days, that it leave not the mouth (of men),*
> *That my name be spread wide in the land,*
> *That I be eulogized in all the lands,*
> *I, the runner, rose in my strength, all set for the course, from Nippur to Ur,*
> *I resolved to traverse as if it were (but a distance) of one "double hour."*
> *Like a lion that wearies not of its virility I arose,*
> *Put a girdle (?) about my loins,*
> *Swung my arms like a dove feverishly fleeing a snake,*
> *Spread wide the knees like an Anzu bird with eyes lifted toward the mountain.*

From the translation by Samuel Noah Kramer, *History Begins at Sumer*, University of Pennsylvania Press, Philadelphia (1981): 286.

Shulgi, on arriving at Ur makes sacrifices in the Temple of Sin accompanied by music. He rests, eats and bathes in his palace and then returns to Nippur during a rain and hail storm. Arriving back in Nippur, he banquets and drinks beer with the Sun-God Utu and Shulgi's divine spouse, the fertility goddess Inanna. In this hymn, clearly Shulgi is not competing with any other runner, nor does he run on a canonical racecourse. He is running on roads which, incidentally, he mentions earlier in the same hymn as being well maintained under his care.

From post-Old Babylonian texts (ca. 1200 B.C.) are references to athletics in connection with various religious festivals, and

---

1. See Åke Sjöberg, "Trials of Strength, Athletics in Mesopotamia," in 'Exploring 5000 Years of Athletics,' David Gilman Romano, ed., *Expedition*, 27, no. 3 (1985): 7-9.

2. Ibid., p. 7.

3. Ibid., p. 8.

4. Samuel Noah Kramer, *History Begins at Sumer*, Philadelphia, 1981. See especially Chapter 31, "Shulgi of Ur: The First Long Distance Champion," pp. 284-288.

tural structures for the contests, several texts refer to the great court-yard of the temple as the site of the contests. The few extant representations of athletics in the form of sculpture and reliefs from Mesopotamia from as early as 2900 B.C. give us no more information about athletic structures.[5]

From the Old Kingdom in Egypt there is evidence for athletics from the site of Saqqara. In ca. 2650 B.C., King Djoser of the Third Dynasty used an architect named Imhotep to design a massive stone tomb for himself. The tomb was ultimately designed in the form of a stepped pyramid with surrounding buildings, an entire precinct of structures, including tombs, temples and altars as well as open spaces which is collectively known today as the Step Pyramid of King Djoser.[6] The entire walled enclosure measures ca. 544 x 277 m. A part of this complex of buildings is what has been identified as a cult racecourse, located on the south side of the step pyramid in the Great South Court (Fig. 7). The structure consists of a broad and flat running surface with two large B–shaped masses of stone 55 m apart. It has been suggested by the excavators that these stone objects may have been altars and it is likely that they also served as turning posts around which the Pharaoh would run.[7]

It was on this ritual racecourse on which King Djoser ran during certain important religious occasions during his lifetime, for instance, at the dedication of the temple precinct and at the 30th anniversary of the King's rule, known as the Heb-Sed Festival. The King apparently did not compete against any other runners during these occasions.[8] From the area of the tomb complex at Saqqara come three similar sculptural reliefs which include an image of King Djoser running in this ceremonial race ca. 2650 B.C. One of these three reliefs is illustrated as Fig. 8. He runs from the right to the left and in his right hand he holds a whip, symbolic of royal rule, and a *mekes* or a rolled document in his left hand.[9] This relief was set up to commemorate 30 years of his rule as Pharaoh. The hieroglyphics on the same relief state that Djoser ran a specific number of lengths of the racecourse.[10]

One additional note should be made of the architectural arrangement of the Great South Court at Saqqara. The western aspect of the court is bounded by what have been interpreted as three successive terraces that are usually thought to be dummy magazines. The terraces rise in elevation toward the west and there may be steps or a ramp that lead up to the highest of these terraces from the northern end of the precinct. Although the function of these terraces is not completely understood, the idea of terraces flanking a cult racecourse suggests the possibility that one or more of the western terraces may have been used as levels from which spectators (if there were any)

---

5. See above note 1, Åke Sjöberg, "Trials of Strength," fig. 1, miniature copper statuette from Khafaji, Iraq, showing two belt wrestlers, ca. 2600 B.C.; fig. 3, stone stela from Badra, Iraq, showing the earliest representation of wrestlers, ca. 2900 B.C.; fig. 5, a clay tablet from a tomb at Sinkara (Larsa) in Southern Iraq showing boxers and musicians, ca. 1200 B.C. See also Henri Frankfort, *More Sculpture From The Diyala Region*, Oriental Institute Publications, Vol. LX, Chicago (1943): 11-13; 15; pls. 54, 62.

6. Cecil M. Firth, J.E. Quibell with plans by J.-P. Lauer, *Excavations at Saqqara, The Step Pyramid*, Cairo, 1935. For a summary of the excavated precinct see I.E.S. Edwards, *The Pyramids of Egypt*, New York (1988): 34-69. A recent consideration of the Step Pyramid of Djoser complex has been made by Barry J. Kemp, *Ancient Egypt, Anatomy of a Civilization*, New York (1989): 55-61. See also David O'Connor, "Boat Graves and Pyramid Origins, New Discoveries at Abydos, Egypt," *Expedition* 33, no. 3 (1991): 5-17, regarding recent discoveries at Abydos relating to the origins of the Step Pyramid at Saqqara. There is a helpful table of Egyptian chronology included on p. 7.

7. See Jean-Philippe Lauer, *Fouilles à Saqqarah, La Pyramìde à Degrés, L'Architecture*, Paris, 1936, Vol. 1: 167-169; 178-179.

8. For a summary of the most recent research of the Festival of Sed and the significance of the run, see Wolfgang Decker, *Sports and Games of Ancient Egypt*, New Haven (1992): 24-34. Decker, p. 29, suggests that the racecourse "was an element of the royal funerary complex and not the stage for the actual run that took place during the Festival of Sed."

9. See above note 6, Firth, Quibell and Lauer, *Excavations at Saqqara*, p. 104, pl. 16.

10. There are also representations of running from the first dynasty in Egypt. See Wolfgang Decker (above note 8): 30-33.

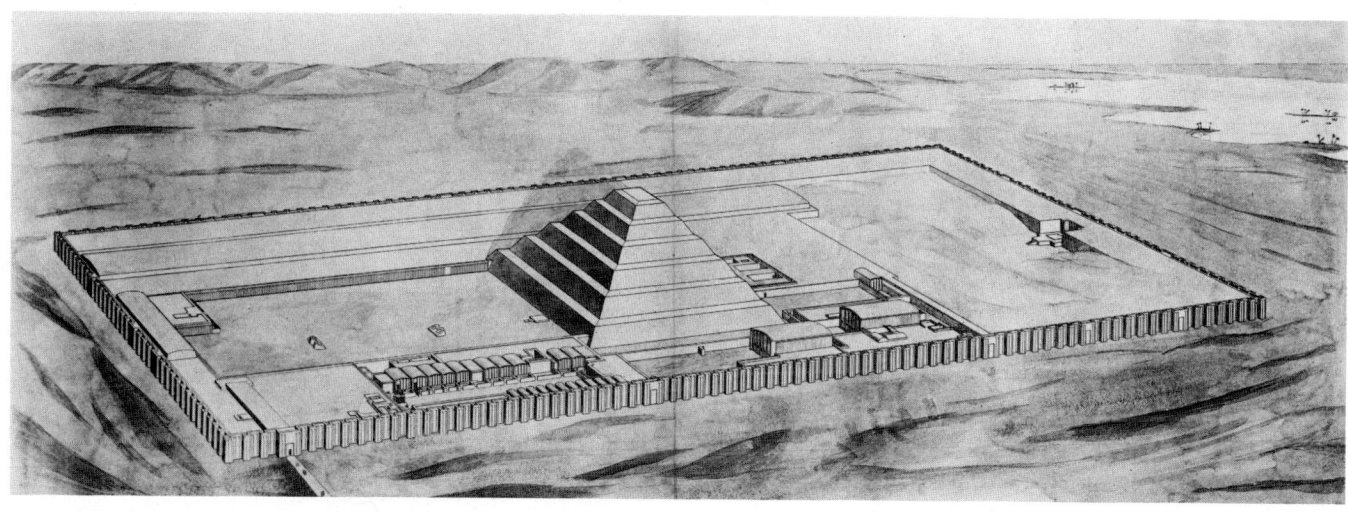

7. Perspective drawing of the Step Pyramid of King Djoser complex at Saqqara, ca. 2650 B.C. View is towards the northwest. The Great South Court is to the left of the Step Pyramid. From Jean Philippe Lauer, La Pyramide à Degrés, L'Architecture, Cairo, 1936.

could have stood to watch the Pharaoh run on the cult racecourse.[11] Since the terraces continue for much of the length of the precinct, could they also have been used for spectators to watch other ceremonies that took place around the step pyramid, to the north as well as to the south?[12]

There are other examples of representations of athletes and athletics from Egypt including another stone relief of a "royal run." From the 18th Dynasty of the New Kingdom, ca. 1480 B.C. is a relief from Karnak depicting Queen Hatshepsut, dressed as a man, running in a cult exercise similar to King Djoser's.[13] It appears that the tradition of the "cult run" continued from the Old Kingdom through the Middle Kingdom and into the New Kingdom, and was practiced in different locations. In addition, scenes of wrestling as well as gymnastics and other activities are documented from the Old Kingdom, from ca. 2300 B.C., the Tomb of the Vizier Ptahhotep at Saqqara.

Thus, from the third millennium B.C., from both Mesopotamia and Egypt, there is evidence suggesting that athletics were held in a religious context in the courtyard of a temple or palace. The specific arrangements of the temple courtyard is not known from Mesopotamia nor do we know if there were contests in events other than wrestling, but we do get a somewhat clearer idea about the arrangements at Saqqara where there is a cult racecourse associated with a tomb, temples and altars. We do not know if these eastern or

11. Wolfgang Decker (above note 8): 29-30, suggests that the sacred buildings located to the east of the racecourse might have symbolically signified the presence of spectators.

12. The only other possible facility for spectator accommodation that I am aware of from ancient Egypt is known from the time of Amenophis III. In the region of Thebes there has been identified a strip of cleared desert, 4.1 km. long and 120 m. wide that has been associated with a possible chariot racecourse, with a viewing facility nearby. See Barry J. Kemp, "A Building of Amenophis III at Kôm El-'Abd," *Journal of Egyptian Archaeology*, 63 (1977): 71-82. I thank David O'Connor for this reference.

13. See A.D. Touny and Steffen Wenig, *Der Sport in Alten Ägypten*, Lausanne (1969): plate 63.

8. Stone relief of King Djoser at Saqqara. South Tomb, northern stela, ca. 2650 B.C. From Jean Philippe Lauer, *La Pyramide à Degrés, L'Architecture*, Cairo, 1936.

"oriental" customs, traditions and facilities were known to the Greeks many centuries later although as will be seen, there are many and varied influences from Egypt and Mesopotamia coming to Greece during the Orientalizing Period in the eighth and seventh centuries B.C. in the realm of science, mathematics, art, architecture, and language.

There are a limited number of literary and epigraphical sources that have to do with athletic competitions in Egyptian contexts from this general period of time. A stone inscription has been discovered fairly recently and has been published, the "Running Stela of Taharqa," dating to the twenty-fifth dynasty, 685-684 B.C.[14] It is the description of a footrace organized by the king Taharqa between soldiers from Memphis through the desert to Fayum and return, a total distance of ca. 100 kilometers. As described, it was a race run in two segments separated by a two-hour interval. The king accompanied the runners in a chariot, and at a certain point dismounted and joined the runners for a segment of the distance.[15] The winner of the race received praise and the king arranged for him to eat and drink with his bodyguard. All of the finishers received unspecified prizes. This competition was, of course, run on desert roads, not on a racecourse, and it formed a part of the soldier's physical conditioning.[16]

There are several passages from the fifth century B.C. Greek historian Herodotus that have to do with Egyptians and Greek athletics. There is the account (2,160) where Herodotus describes an assembly of ambassadors from Elis going to visit Psammis, the king of Egypt in the early sixth century B.C. The Greek ambassadors were seeking advice and criticism about the organization and running of the Olympic Games. There is also the account (2,91) where Herodotus says that Egyptians do not adopt Greek customs or any foreign customs at all with the exception of those living near Khemmis, near Thebes, where like in Greece, they celebrate contests of all kinds in honor of Perseus, in which livestock, cloaks and skins are awarded as prizes.

Literary, Epigraphical and Historical Evidence

Athletic stories and references are a part of ancient Greek literature as early as the *Iliad* and *Odyssey* of Homer that are commonly understood to have been written down in the years around 700 B.C.[17] When referring to the place of human athletic contests, Homer uses the word *dromos* from the verb δραμεῖν, to run, meaning "the running place," or "racecourse" as in the *Odyssey,* 8,121, where the Phaeacians provide athletic contests in their place of assembly for the visitor Odysseus. So too, in the *Iliad,* 23,758, a very modest *dromos* is marked out for the five contestants at the funeral games of Patroklos on the plain of Troy.[18]

14. See Wolfgang Decker (above note 8): 62-66. See also Ahmed M. Moussa, "A Stela of Taharqa from the Desert Road at Dahshur," *Mitteilungen des Deutschen Archäologischen Instituts, Abteilung Kairo*, 37 (1981): 331-337; Hartwig Altenmüller and Ahmed M. Moussa, "Die Inschriften auf der Taharkastele von der Dahschurstrasse," *Studien zur Altägyptischen Kultur* 9 (1981): 57-84; W. Decker, "Die Lauf-Stele des Königs Taharka," *Kölner Beiträge zur Sportwissenschaft* 13 (1984): 7-37.

15. This in itself is reminiscent of the ceremonial equestrian contest known from Greek athletics as the *apobates* (ἀποβάτης). See Donald G. Kyle, *Athletics in Ancient Athens*, Leiden (1987): 188-189.

16. There are some similarities between this account and the description of Shulgi's run between Nippur and Ur, above p. 9. Both are run over extremely long distances and both include an interval between the two parts of the run. Although unspecified in the hymn, Shulgi's run (or a portion of it) may have been at night.

17. See Cecil Maurice Bowra, "Homer," in *The Oxford Classical Dictionary*, Second Edition, Oxford (1970): 524-526.

18. In both instances, *Iliad* 23, 758 and *Odyssey* 8, 121, the same phrase is used, τοῖσι δ'ἀπὸ νύσσης τέτατο δρόμος, "a racecourse was marked out for them from the starting point."

The earliest use of the word *stadion* in Greek literature is found in the early fifth century in the extant works of Simonides, Pindar and Bacchylides, where there are multiple references to the *stadion* as a structure, a race and to *stadiodromos* runners.[19]

Later in the same century, both Herodotus and Thucydides use the word *stadion* as a measure of distance. It is Herodotus (2,149) who defines the *stadion* as the equivalent of 6 *plethra*, 100 *orguiai*, or 600 feet.

Possibly the earliest extant appearance of the word *stadion* is found on a Panathenaic amphora now in the Metropolitan Museum of Art (1978.11.13). On the obverse is portrayed the archaic, armed Athena, striding to the left; the reverse depicts three sprinting runners. The athletic scene includes a painted inscription giving the name of the contest in which the amphora was won.[20] It reads ἀνδρῶν σταδιό, "of the *stadion* race for men." The amphora is dated to the mid-sixth century B.C. (Fig. 9a, 9b) and as such would pre-date the first literary use of the word by at least 20-30 years. The vessel is also important since it is a so-called "prize amphora," including on the obverse the painted inscription, τῶν ἀθενήθεν ἄθλων, "of the contests at Athens." The painted inscription includes the name of the manufacturer (potter) of the vessel, Νίκιας ἐποιεσεν, "Nikias made (me)."[21]

Although the etymology of the word *stadion* is obscure, it has been suggested that its origin may derive from the Greek verb, ἵστημι, "to stand."[22] There is substantial evidence to suggest (discussion below) that the earliest artificial embankments built along the sides of racecourses or *dromoi* were used as areas on which spectators stood, and this author has suggested elsewhere that the name of the facility *stadion* may have derived from the use of the embankments, and that the word *stadion* may originally have meant "the standing place."[23]

It seems possible, therefore, that the derivation of the word *stadion* may come from the verb ἵστημι, "to stand" and that the earliest meaning of the word was "the standing place." Homer did not use the word στάδιον but he did use the word στάδιος, meaning "standing fast and firm." If, in fact, the word *stadion* does come from the verb 'to stand' we should consider the implications for the other two uses of the word, as a distance and as a footrace. As discussed above, the earliest use of the word *stadion* in literature is found in Simonides, Pindar and Bacchylides, where there are multiple references to the *stadion* as a structure and footrace and to *stadiodromos* runners.[24] The word referring to the race occurs earlier on a mid-sixth century B.C. Panathenaic amphora, predating the earliest literary use of the word by at least 20-30 years. Since Herodotus is the first to define the

---

19. Simonides 519, frags. 92.3, 96.3, 99.2, 85.4 (D.L. Page, *Poetae Melici Graeci*, Oxford, 1962). Pindar, *Olympia* X, 64; XIII, 30; XIII, 37; *Pythia* XI, 49; *Nemea* VIII, 16; *Isthmia* I, 23. Bacchylides VI, 7; VI, 15; IX, 21. There is also a single use of the word *stadion* in Theognis although this passage, 1306, is thought to have been a later addition.

20. ABV 408.1. The amphora has been most recently published in *Notable Acquisitions 1975-1979, The Metropolitan Museum of Art*, selected by Phillipe de Montebello, New York, 1979. The entry is by Dietrich von Bothmer, p. 14.

21. Georg von Brauchitsch, *Die Panathenaischen Preisamphoren*, Leipzig and Berlin (1910):11, cites another Panathenaic amphora, Munich 498, dated to the mid-sixth century B.C. which has a painted inscription, σταδιο ανδρων

νικε, "winner of the men's *stadion*," which occurs on the reverse of the amphora above four sprinting runners. A facsimile of the scene is found in E.N. Gardiner's *Greek Athletic Sports and Festivals*, London (1910): 281, fig. 52.

22. Pierre Chantraine, *Dictionnaire étymologique de la langue grecque*, Paris (1968): 1041. Chantraine suggests two possibilities, from σπάω, and from στάδιος (ἵστημι). Hjalmar Frisk, *Griechisches etymologisches Wörterbuch*, Vol. 2, Heidelberg (1970): 773, believes that the word *stadion* derives from σπάδιον by dissimilation.

23. See Romano, "Stadia" (1981): 241-276 and Romano, "Arete" (1983): 9-16.

24. H. Frisk (above note 22) believes that the earliest use of the word is as a racecourse, and the second use is as a measure of length.

9a. Panathenaic Amphora (Metropolitan Museum of Art
1978.11.13), ca. 550 B.C. Obverse, showing striding Athena.
Photos courtesy of the Metropolitan Museum of Art.

9b. Reverse, showing three sprinting athletes.

word *stadion* as a measure of distance, it would seem a good possibility that the word *stadion* came to be used first for the area where spectators stood bordering the *dromos* on low embankments; later the word was applied to the name of the athletic event held there; and finally the term was applied to the specific length of the racecourse.[25] By association, the word came to mean the combination of the spectator area and the competition area.

There are very few seat blocks found in Archaic and Classical stadia and these seem to have been reserved for judges and VIP's.[26] It was not until the Hellenistic period that seats became fairly commonplace in Greek stadia although by no means necessary. By the fifth century B.C., therefore, the Greek word *stadion* had come to have three meanings: a structure for athletic contests; a footrace one length of the structure; and a linear distance always equal to 600 feet.

The word *dromos* could refer to the racecourse of a *stadion* or to a facility without any formal accommodation for spectators, in which case it would serve only as a racecourse. Examples of this variety of *dromos* are found in the gymnasium at Olympia where parallel *dromoi* have been found in the eastern stoa of the Hellenistic gymnasium, north of the *palaistra*.[27] Fragments of stone starting line blocks were found near the east stoa and would originally have been located at the south and the north ends of the stoa, as a *xystos,* 192 meters apart. The distance was the equal of the length of the *dromos* of the contemporary Olympia stadium. Other examples of *dromoi* from the Peloponnesos are known from literary evidence to have existed at Elis, Sikyon and Sparta and they must have been common elsewhere (Fig. 3).[28] The description of Pausanias of the *dromos* from Sparta is especially interesting in comparison to Corinth since the account is of a racecourse where the young men practice running in the middle of the city with shrines and sanctuaries surrounding it. At Elis, there is a *dromos* for the runners and pentathletes to practice on as well as a *hieros dromos*, a sacred racecourse, for the competing runners.

We learn from Pausanias that a *stadion* and a *dromos* are not the same and, in fact, have several important differences. A *stadion*, according to the descriptions of Pausanias, is a facility for spectators, either in the form of banked earth or stone seats, which is constructed along the sides of a racecourse.[29]

A *dromos* is a racecourse, proper, without formal spectator facilities, which can either be a place for athletes to train or a place for athletes to compete. Three inscriptions of the mid-sixth century B.C., found on the Athenian Akropolis, include the word *dromos*.[30] According to Raubitschek, *dromos* in these instances refers to the early contests of the Panathenaic Games. Travlos, on the other hand, believes that the word *dromos* in the same inscriptions refers to the racecourse for the contests.[31]

---

25. The standard view (Liddell and Scott, *A Greek-English Lexicon*, Oxford (1940): 1631) is that the primary meaning of the word is as a standard of length and that the meanings of the word as racecourse and contest are secondary.

26. Romano, "Stadia" (1981): 218-221.

27. P. Graef, "Das Gymnasion," in Curtius and Adler, eds., *Olympia, Die Ergebnisse der von dem deutschen Reich veranstalteten Ausgrabung*, Vol. II: *Die Baudenkmaler von Olympia*, Berlin (1892): 127-128, plate 78; Emil Kunze and Hans Schleif, "Das Gymnasion," *OlBer* III, Berlin (1941): 67-70, plate 6.

28. Elis: Pausanias (VI,23,1-2); Sikyon: Herodotus (6,126) and Sparta: Pausanias (III,14,6-8). See also Romano, "Stadia" (1981): 197-200.

29. Pausanias (II,27,5) tells us that most stadia of Greece are made of banks of earth, γῆς χώματα, and he describes individually the stadia of Epidauros (II, 27, 5); Olympia (VI, 20, 8); Thebes (IX,23,1); Aegina (II, 29, 11); Athens (I,19,6); Delphi (X,32,1) and Isthmia (II,1,7). In each case, the structure that Pausanias describes is not the racecourse floor where the athletes compete, but rather the embankments, whether of earth or stone, built for the spectators.

30. A.E. Raubitschek, *Dedications From the Athenian Akropolis*, Cambridge, Mass. (1949): 350-358, nos. 326, 327 and 328. All three inscriptions begin with τὸν δρόμον ἐποίεσαν τῆι θεõι.

31. John Travlos, *Pictorial Dictionary of Athens*, London (1971): 2. See also the discussion in Donald G. Kyle (above note 15) pp. 26-28. An inscription similar to those in Athens, from Eleusis, *I.G.* I² 817 also mentions the word *dromos*.

Archaeological Evidence

The ancient *stadion* was in its earliest form, a flat, usually rectangular space 600 feet long and 50 to 100 feet wide and was often bordered by natural or artificial earth embankments for the accommodation of spectators. Although the length of the *dromos* of a *stadion* was always 600 feet long by definition of the word, the length of the ancient foot varied considerably from place to place, from structure to structure, as well as from century to century. Therefore absolute *stadion* lengths varied correspondingly. For example, as a result of archaeological excavation, it is known that the *stadion* at Halieis measures 166.50 m. between the front grooves of the starting lines, whereas the *stadion* at Olympia (Olympia III Stadium), measured between the same two points, is 192.28 m., for a net difference of 25.78 m. The resulting foot measures are 0.278 m. for Halieis and 0.320 m. for Olympia.[32]

Limited remains of the artificial earth embankments have been discovered in the major Panhellenic sanctuaries of Olympia and Isthmia, dating from the mid-sixth century B.C. and, as such, constitute the earliest physical remains of the ancient Greek *stadion*.

OLYMPIA

The Sanctuary of Zeus at Olympia, in the western Peloponnesos, was the home of the oldest and the most important sanctuary of Zeus in the Greek world. The traditional date of the foundation of the athletic contests at Olympia is 776 B.C. based on the *Olympic Register* compiled originally by Hippias of Elis ca. 400 B.C. and later worked on by Aristotle and other authors.[33] From differing interpretations of the available archaeological evidence, a date earlier than 776, as well as a later date has been suggested for the original introduction of the contests.[34]

Throughout the history of Olympia, the successive stadia have been located to the east of the *altis*, the cult center of the sanctuary where the Temple of Hera is known from ca. 600 B.C. and the Temple of Zeus from ca. 470 B.C. (Fig. 10). The ash altar of Zeus should date earlier still. Our knowledge of the earliest of the three stadia, Olympia I Stadium, is based upon an Archaic retaining wall and an artificial embankment which must have bordered the Archaic *dromos* to the south.[35] The stone retaining wall, 7.5 m. long and 2.57 m. high runs approximately parallel to the latest of the three chrono-

32. Romano, "Stadia" (1981): 250-267, and Romano, "Arete" (1983): 9-16.

33. See L. Moretti, *Olympioniki. I vincitori negli antichi agoni Olimpici*, Rome (1957): 59.

34. See Alfred Mallwitz, "Cult and Competition Locations at Olympia," in Raschke (1988): 79-109; Hugh M. Lee, "The 'First' Olympic Games of 776 B.C.," *op.cit.*, 110-118; Catherine Morgan, *Athletes and Oracles, The Transformation of Olympia and Delphi in the Eighth Century B.C.*, Cambridge (1990): 47-49.

35. Since the nineteenth century, the Deutsches Archäologisches Institut has been working at Olympia. Their reports on the excavation of the stadium

include the following: Emil Kunze and Hans Schleif, "Das Stadion," *OlBer* II, Berlin (1938): 8-12; Emil Kunze and Hans Schleif, "Das Stadion," *OlBer* III, Berlin (1941): 10-12; Emil Kunze, "Das Stadion," *OlBer* V, Berlin (1956): 1; Alfred Mallwitz, "Das Stadion." *Olber* VIII, Berlin (1967): 16-82. See also Alfred Mallwitz, *Olympia und seine Bauten*, Munich (1972): 180-186 and Wolf Koenigs, "Stadion III und Echohalle," *OlBer* X, Berlin (1981): 353-369. For a summary see Romano, "Stadia" (1981): 115-141. Most recently see Jürgen Schilbach, "Olympia, die Entwicklungsphasen des Stadions," *Proceedings of an International Symposium on the Olympic Games (1988)*, William Coulson and Helmut Kyrieleis, eds., Athens (1992): 33-37.

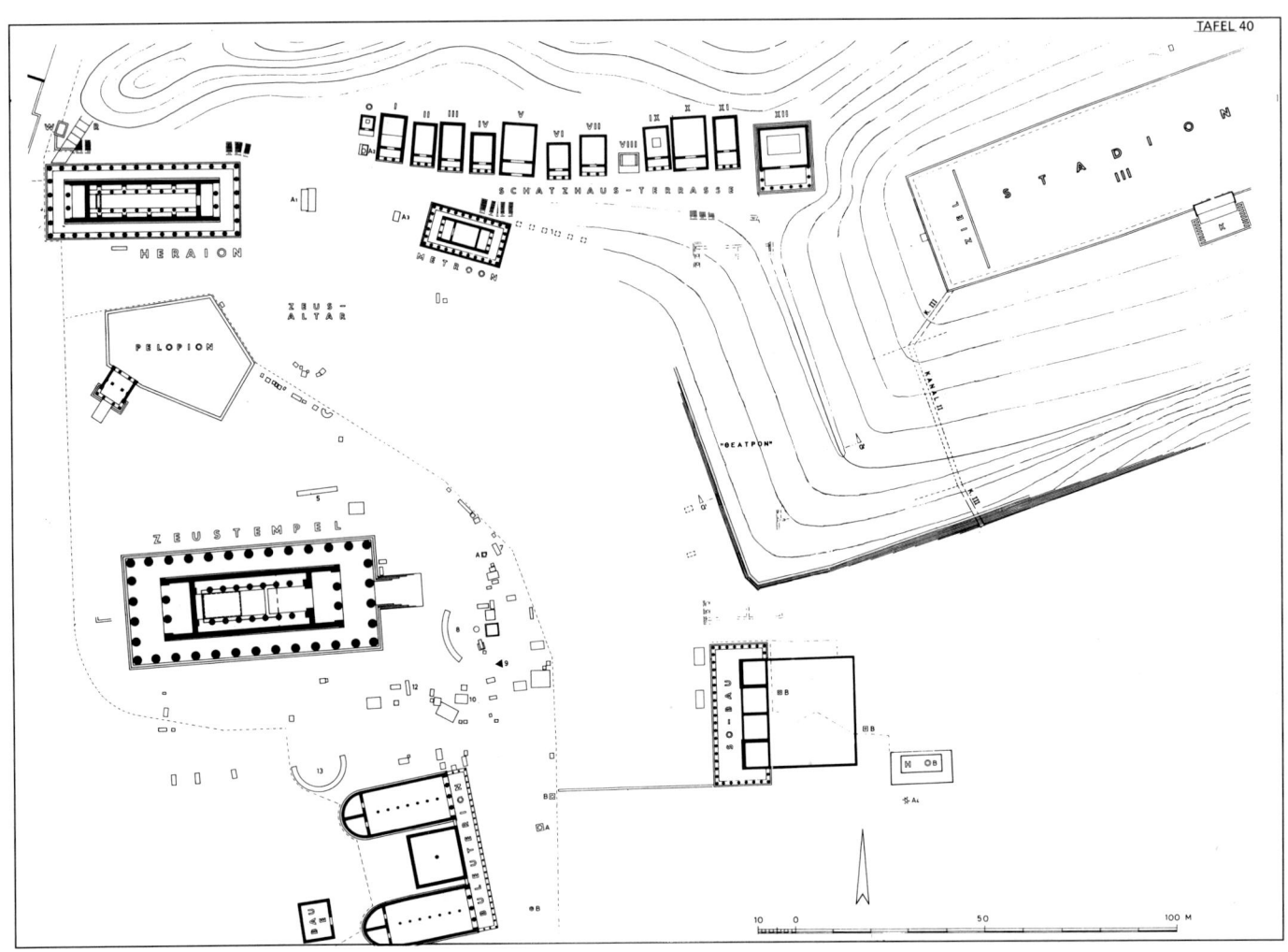

10. Olympia. Plan of the East *altis* and Olympia III Stadium by the late Fifth Century B.C. Courtesy DAI, Athens.

logical phases of the Olympia Stadium (Fig. 11). The wall supported an artificial embankment of earth which sloped down towards the racecourse floor, nothing of which remains since the racecourses of the two succeeding stadia were dug below the earliest racecourse level. It is assumed, however, that the *dromos* of the earliest stadium was open to the west and extended well within the *altis*.[36] No artificial embankment existed on the north side of the Archaic stadium. The date of the fill of the Archaic embankment to the south, based on ceramic evidence, is ca. 540 B.C. It must be assumed that earlier racecourse(s) at Olympia were probably simpler still, possibly a *dromos*, with little or nothing in the way of formal spectator accommodations, e.g., artificial embankments.

Alfred Mallwitz has made the interesting observation that the numerous wells that were dug, on what was the natural lower slope of the Kronos hill, immediately to the north of the Olympia I and II Stadium racecourses, between ca. 700 B.C. and the second quarter of the fifth century B.C., likely represent the areas where the spectators, from the late eighth century B.C. on, gathered to watch the contests in the so-called "Urstadium," what would be the same area as the later *dromos* for the architecturally attested Olympia I Stadium.[37]

Two stone seats can be safely associated with the Olympia I Stadium; both are seats of *proxenoi*, "Gorgos" and "Euwanios," who were local diplomatic representatives of foreign city-states. Each block is dated independently, on the basis of letter forms, to the mid-sixth century B.C. The first is a block of blue-black marble 0.27 X 0.42 X 0.31 m. On the top surface of the block and around its perimeter reads Γόργος Λακεδαιμόνιος πρόξενος ϝαλείων, "Gorgos the Lacedaimonian, Consul of the Eleans," (Fig. 12).[38] The block was found in Roman fill to the south of the Olympia III Stadium. A second seat block was found, built into the southern retaining wall of the eastern open-air approach to the vault of the Olympia III Stadium. It reads Εὐϝάνιου Λακεδαιμόνιοῦ πρόξενοῦ, "of Euwanios the Lacedaimonion, Consul."[39] The top face of the block measures 0.36 x 0.25 m. although the block appears to have been shaved down on one of the sides. The inscription may originally have continued around the fourth side of the block. Both "Gorgos" and "Euwanios" were dignitaries and it is generally understood that stone seat blocks in the stadium, like theirs, were not provided for the common spectators.

Considerably more is known about the Olympia II Stadium which is now dated to the late sixth century B.C.[40] The outline of the *dromos* is known from the trenches left after the robbing of parallel

36. It is likely that the *dromos* of the Olympia I Stadium and that of the Olympia II Stadium were in the same location, although the Olympia II Stadium would have been lower in elevation. See Alfred Mallwitz, "Cult and Competition Locations at Olympia,"in Raschke (1988): 94. Five stone starting blocks were found in 1979 by Alfred Mallwitz, to the south and southeast of the odeon, and are associated with the Olympia I Stadium. See Hector Catling, "Archaeology in Greece," *Archaeological Reports for 1980-1981*, British School at Athens, London (1981): 21.

37. A. Mallwitz, "Cult and Competition Locations at Olympia," in Raschke (1988): 98-99, "the positions of the wells, then, indicate that the oldest setting for the contests, the Urstadium, must have lain where Stadium I and its successors have left traces since the mid-sixth century." Also see Werner Gauer, "Die Tongefässe aus den Brunnen unterm Stadion-Nordwall und im Südost-Gebiet," *Olympische Forschungen* VIII, Emil Kunze and Alfred Mallwitz, eds., Berlin, 1975.

38. Alfred Mallwitz, *Olympia und seine Bauten*, Munich (1972): 184-185. It has been observed that the marble for the "Gorgos" seat block comes from Lakonia and the suggestion made that Gorgos had brought the seat block with him from Sparta (E. Kunze and H. Schleif, *OlBer* V (1944): 164-166, fig. 10, pl. 67.) Dina Peppa-Delmousou notes that the letters on the seat block are of the Archaic Lakonian alphabet, dating to 600-550 B.C. (*Mind and Body*, 1989, no. 110, pp. 219-220.)

39. Alfred Mallwitz, "Zu den Arbeiten im Heiligtum von Olympia während der Jahre 1967-1971," *ArchDelt* 27 (1972) *Chronika*: 273-276, plates 211-212.

40. A. Mallwitz "Cult and Competition Locations in Olympia," in Raschke (1988): 94, n. 67. Wolf Koenigs, "Stadion III und Echohalle," *OlBer* X, Berlin (1981): 353-369 dates the construction of the Olympia II Stadium to ca. 470 B.C.

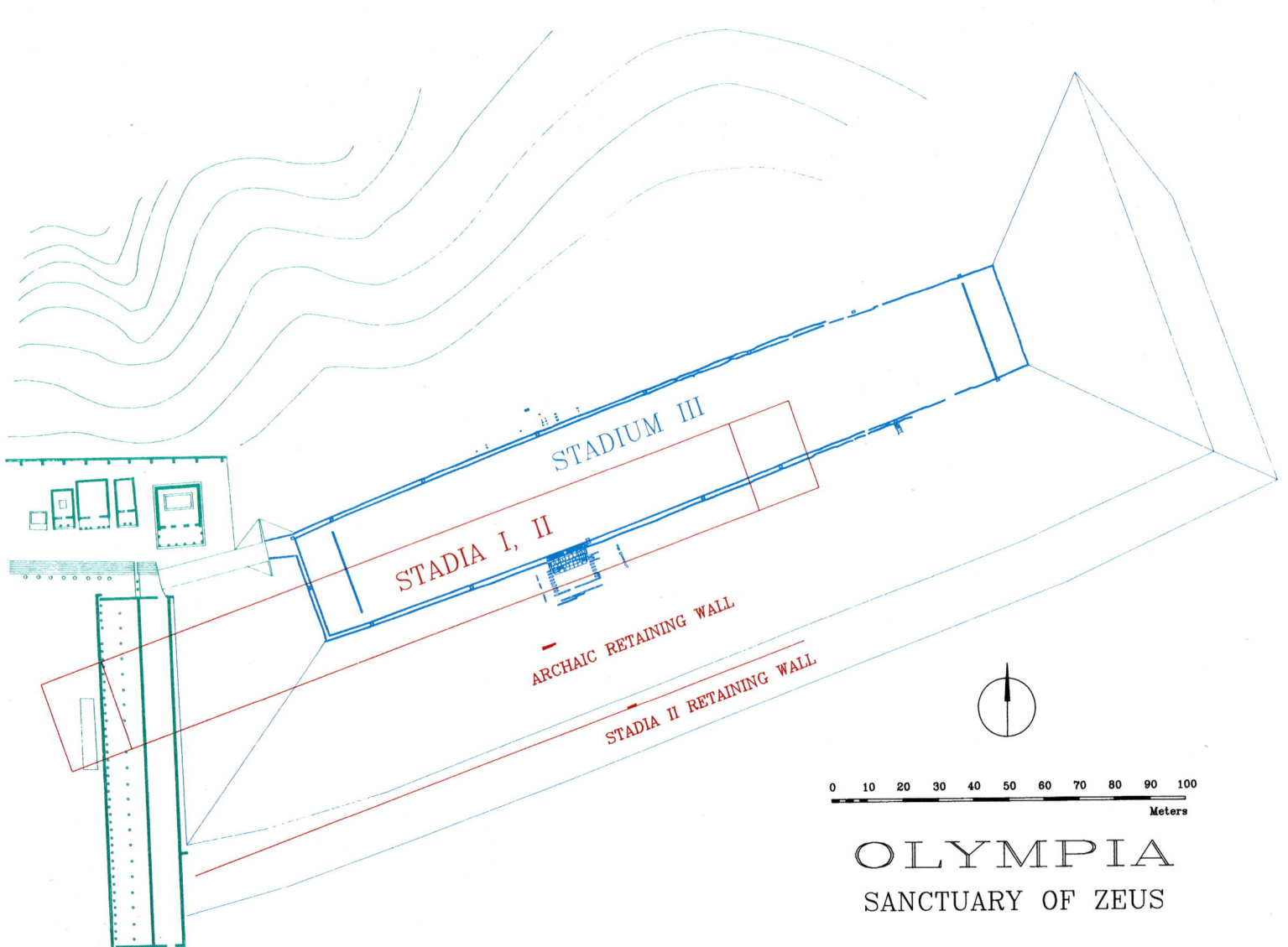

STADIUM III

STADIA I, II

ARCHAIC RETAINING WALL

STADIA II RETAINING WALL

0 10 20 30 40 50 60 70 80 90 100

Meters

OLYMPIA

SANCTUARY OF ZEUS

11. Olympia. Relative locations of Olympia I Stadium and Olympia II Stadium (red) and Olympia III Stadium (blue). Archaic retaining wall of south embankment of Stadium I, ca. 540 B.C., and Stadium II retaining wall, are visible to the south of the *dromos*. Drawing by Gus Fahey after the plan by the DAI, Athens.

12. Olympia. *Gorgos* seat block with inscription, ca. 550 B.C. Photo by the author.

elements, probably curb stones, bordering the north and south edges of the *dromos* from its western starting line. The Olympia II Stadium has a rectangular *dromos*, approximately 26 m. wide, although its length has not been determined by direct measurement, since no trace of its eastern starting line was found.[41] Overrun space exists at the western end of the racecourse for a distance of approximately 19 meters. It is clear that the Olympia II stadium extended well within the *altis* (Fig. 11). The western starting line lies due south of the treasuries X and XI, of Metapontum and Gela, and due north of the Southeast Building and only 100 meters from the northeast corner of the Temple of Zeus. The racecourse floor is described as greenish-yellow in color and as having a composition of clay and sand.[42]

Artificial spectator embankments were created for the Olympia II Stadium on both its north and south sides but there is no evidence for spectator embankments to the east or west. The northern embankment did not require a retaining wall, but rather used the natural contours of the Kronos hill. The artificial fill covered the numerous wells on the hillside which had been in the area during the Archaic period.[43] The slope of the northern embankment was 1:6 or 9 degrees. The southern embankment did require a retaining wall, and it survives in part at a distance of approximately 38 m. south of

41. I have suggested that the 600 foot length of the *dromos* of the stadia at Olympia, before the construction of the Temple of Zeus, may have been measured according to a different foot length resulting in a shorter absolute distance. See Romano, "Stadia" (1981): 255-256, note 23 and Romano, "Arete" (1983): 9-16.

42. Four starting line blocks composed of shelly limestone were found to the south of the Olympia III Stadium built into a drainage canal "K$_\omega$." The blocks were found as a part of the pre-Roman construction and are characterized as having two parallel grooves. The German excavators have associated these four blocks with the Olympia II Stadium, although they may be more likely associated with the Olympia III Stadium, based on their typology. See Emil Kunze, "Das Stadion," *OlBer* V (1956): 17, fig. 4.

43. See above note 37.

22

and parallel to the robbing trench of the south "curb" of the Olympia II Stadium (Fig. 11). Enough of the southern embankment exists to ascertain that the slope of the bank varied from 1:12 or 4 degrees to 1:9 or 7 degrees. I have estimated the spectator accommodation of the two embankments to be approximately 24,000.[44] The preserved section of the wall is stepped in design and is 3.5 m. in height providing an embankment sloping down from south to north toward the racecourse. There is no evidence for the presence of seats on either the northern or the southern embankments.

The Olympia III Stadium replaced the Olympia II Stadium, being approximately 10 meters north and 75 meters east of the former (Figs. 11, 13). There were three major phases of construction of this stadium, during its long use of almost 900 years. The stadium is characterized as having a *dromos* with slightly curved convex long sides on the north and the south, with straight short sides on the east and west. Its racecourse floor was slightly lower than that of the Olympia II Stadium. The artificial spectator facilities were redesigned and enlarged from those of the Olympia II Stadium and, in addition, eastern and western embankments were added. The slope of the southern embankment was 1:9 or 7 degrees and the slope of the northern embankment was 1:6 or 9 degrees (Fig. 13).

The area reserved for the judges, *Hellanodikai,* as described by Pausanias (VI,20,8-9) is found on the south side of the *dromos* at approximately 1/3 of the length of the track from the west end. The judges facilities were enlarged on a number of occasions, to include chairs, a platform and benches. Pausanias explains that opposite the facilities for the *Hellanodikai,* on the north embankment of the stadium, was the altar of the Priestess of Demeter Chamyne. Fragments of the marble altar have been found, as well as the remains of five rows of wooden benches below the altar. There appear to have been no other provisions made for seats in the Olympia III Stadium and, therefore, it must be assumed that the spectators sat or stood directly on the artificial earth embankments. Pausanias also mentions that, according to the Eleans, the tomb of Endymion is at the end of the stadium where the starting place for the *stadion* runners is.[45] I have estimated the spectator capacity of the Olympia III Stadium to be approximately 43,000 or roughly double the capacity of the Olympia II Stadium.

The initial construction date of the Olympia III Stadium must be before the building of the Echo Colonnade and is now suggested to be at the same time as the construction of the Temple of Zeus in the Early Classical period. Before the construction of the Echo Colonnade, the west side of the western embankment sloped down toward the *altis* and was retained by a wall similar to the Treasury Terrace retaining wall (fig. 10). This slope to the west, which was approximately 8 degrees, has recently been interpreted by Koenigs as the *theatron* that Xenophon (VII,4,31) mentions in his account of the battle between the Eleans and the Arkadians in 364 B.C. during the 104th Olympic Games.[46] The slope of the *theatron* must have been used by spectators to watch the wrestling event of the *pentathlon* that Xenophon describes as having been held in the space between the *dromos* and the altar.

44. Here, as elsewhere, I have used the general principle that each spectator would have occupied 0.5 square meter of space on the embankment (standing room) or 0.5 square meter of space on a bench type seat, where applicable.

45. This is usually interpreted as meaning the eastern end of the racecourse.

46. Wolf Koenigs, "Stadion III und Echohalle," *OlBer* X, Berlin (1981): 353 ff., pl. 39. See also A. Mallwitz, "Cult and Competition Locations at Olympia," in Raschke (1988): 94. Jürgen Schilbach (above note 35) pp. 34-35, figs. 3, 5, suggests that the "theatron" is further to the north, backing against the inclination of the treasury terrace.

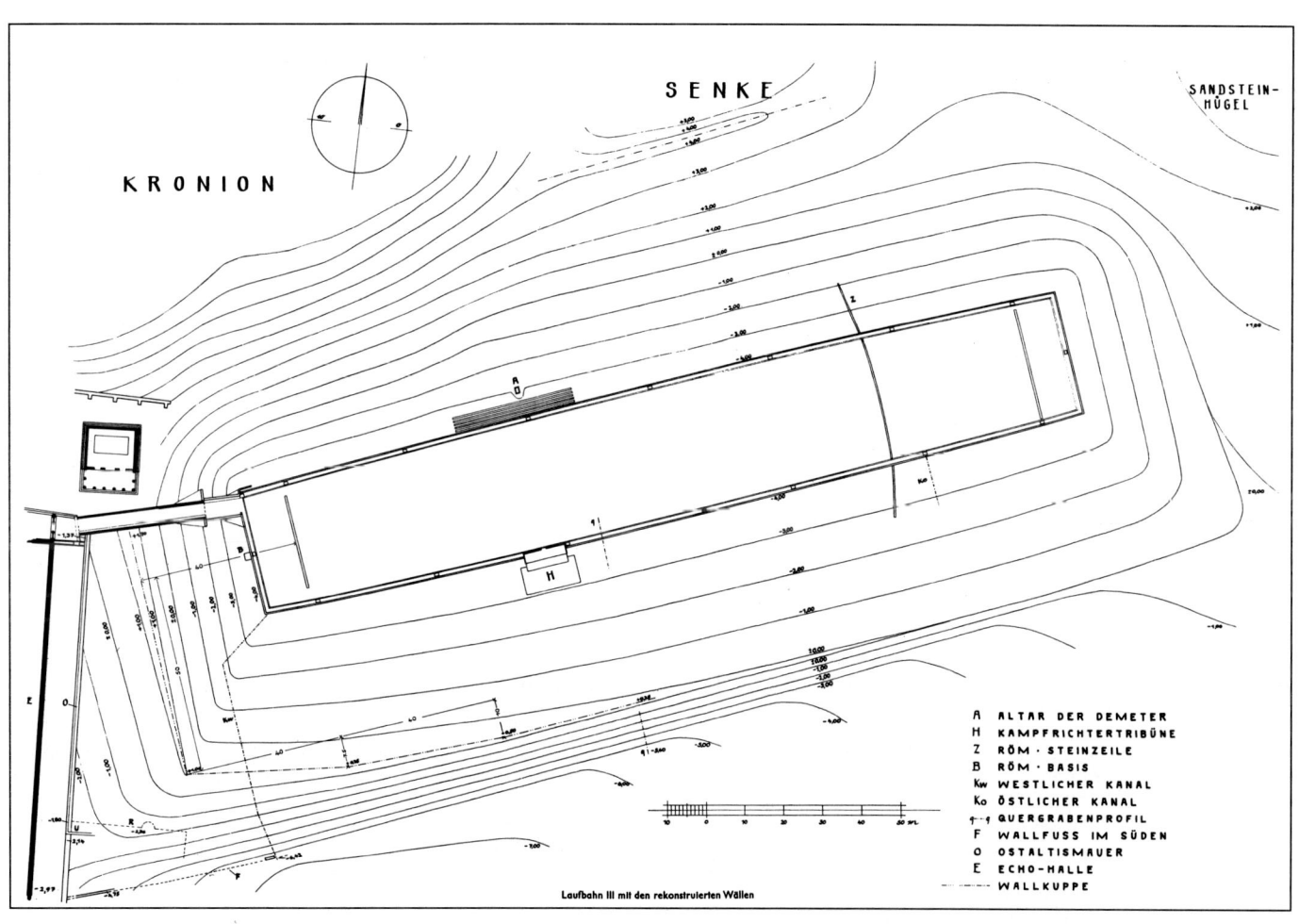

SENKE

KRONION

SANDSTEIN-HÜGEL

A   ALTAR DER DEMETER
H   KAMPFRICHTERTRIBÜNE
Z   RÖM · STEINZEILE
B   RÖM · BASIS
Kw WESTLICHER KANAL
Ko ÖSTLICHER KANAL
⊢⊣ QUERGRABENPROFIL
F   WALLFUSS IM SÜDEN
O   OSTALTISMAUER
E   ECHO-HALLE
------ WALLKUPPE

Laufbahn III mit den rekonstruierten Wällen

13. Olympia III Stadium with reconstructed embankments. Plan courtesy DAI, Athens.

There were a number of renovations of the Olympia III Stadium during the Hellenistic and the Roman periods. The reconstructed stadium, as it appears today, is based on its general condition in the second century A.D. (Fig. 4). At the northwestern corner of the Olympia III Stadium is a vaulted entrance, 32.10 m. long with an eastern approach open to the air, 9.50 m. long. The vault appears to have been a later addition to the stadium, possibly as late as the second century B.C. (Fig. 5).[47] The eastern and western starting lines were found *in situ*, composed of hard limestone blocks. They are found 192.28 m. apart (measured from the forward groove of each line of blocks.) The starting line blocks have two parallel grooves on their top surfaces interrupted by square post holes.[48] The maximum preserved length of the eastern starting line is 28.26 m and of the western starting line is 24.00 m.[49] The distance between post holes and the resulting width of the starting positions vary considerably. There exist 20 or 22 starting positions in the eastern starting line and 18 starting positions in the western starting line. Most of the post holes are 0.04 to 0.08 m. square although several are considerably larger. The two grooves are each approximately 0.045 m. wide and have a beveled forward wall and a vertical back wall. The grooves are approximately 0.16 m. apart, measured center to center. It is likely

that these starting line blocks were a later replacement for earlier blocks.[50] Hydraulic facilities, although not original to the Olympia III Stadium, provided water in stone water channels on all four sides of the stadium. The water originated from the northwest and the water channel blocks carried the water around all sides of the *dromos* to water basins. To the outside of the water channel was found a stone curb.

A curved line of straight stones, approximately 60 m. long and 0.95 m. wide, dating to the Roman period, lies within the floor of the racecourse and extends to the north and south on the slopes of the northern and southern embankments (Fig. 13). It has been suggested that the stones indicate the starting line of the girls' footraces in honor of Hera that is described by Pausanias (V,16,2).[51]

Isthmia

The Sanctuary of Poseidon at Isthmia was a Panhellenic sanctuary located only eleven kilometers east of the ancient city of Corinth. The sanctuary was, throughout most of its history, under the administration of the city of Corinth. Two partially excavated stadia

47. For a recent study of this question see Wolf-Dieter Heilmeyer, "Durchgang, Krypte, Denkmal: Zur Geschichte des Stadioneingangs in Olympia," *AM* 99 (1984): 251-263. Heilmeyer suggests that the vault was constructed in the second half of the second century B.C.

48. Alfred Mallwitz, "Das Stadion," *OlBer* VIII (1967) plate 9. Stone starting blocks with two parallel grooves have been found in other Peloponnesian stadia; Epidauros, Halieis, Nemea, the Later Stadium at Isthmia as well as the Olympia Gymnasium *xystos*. Individual starting line blocks have been found at Olympia, Tegea, Mt. Lykaion, and Argos. See Romano, "Stadia" (1981): 205-210. Starting lines with two parallel grooves have been found outside the Peloponnesos in the Delphi Stadium (P. Aupert, "Le Stade," *FdD*, II, *Topographie et Architecture*, part 7, Paris, 1979.)

49. The width of the *dromos* of the Olympia III Stadium, as measured between

the inside faces of the foundation curbs at the western starting line is 31.54 meters and between the restored foundation curbs at the eastern starting line is 31.26 m.

50. See above note 42 . Three stone blocks from the Roman parapet of the *Hellanodikai* area, blocks E, F, G (*OlBer* III (1939): 10-12, fig. 5), above note 35, appear to be blocks which agree generally in size and shape with blocks associated with *huspleges* at other stadia: the Later Stadium at Isthmia, Epidauros, Nemea and the Hellenistic *dromos* in Corinth. These blocks, coming from an earlier phase of the Olympia III Stadium, may have been reused in the Roman period in the *Hellanodikai* construction.

51. Alfred Mallwitz, "Das Stadion," *OlBer* VIII, Berlin (1967): 60. See below Chapter 6, p. 96 note 8.

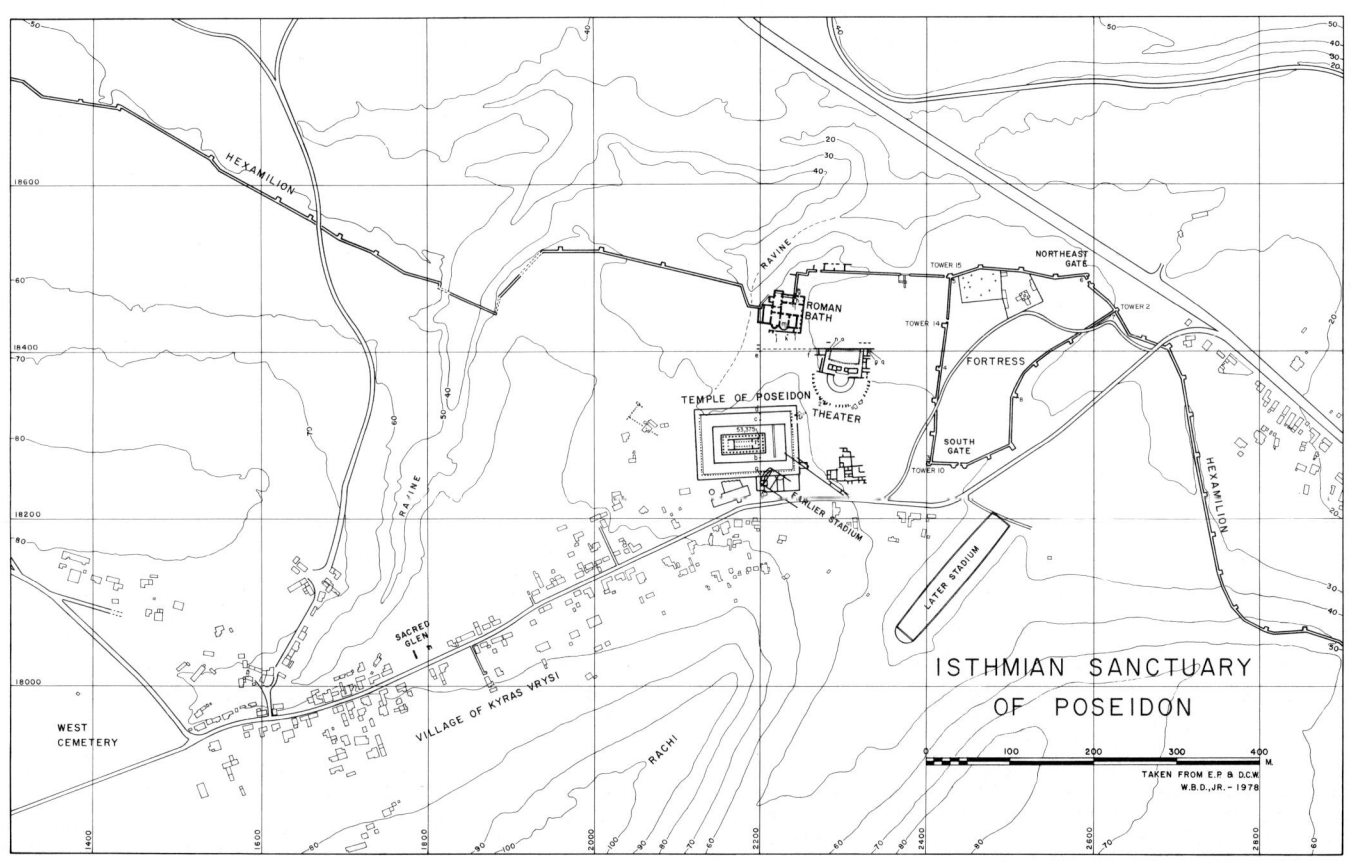

14. Isthmia, Sanctuary of Poseidon. Plan courtesy of The Ohio State University, Isthmia Excavations.

are known at Isthmia, and are known as the Earlier and the Later Stadium.[52] The Earlier Stadium is located immediately adjacent to the site of the Archaic and Classical temples of Poseidon; the Later Stadium is located about 240 m. to the southwest of the temple (Fig. 14).[53]

The evidence for the Archaic phase of the Earlier stadium is evidenced only by a level of hard stone packing to the northeast of the stadium floor of the Classical phase.[54] The stone packing is clearly evident in three separate areas, each of which is approximately 10-15 m. north and east of the east edge of the racecourse of the Classical phase and roughly parallel to it (Fig. 15). The outer edge of the stones is larger and more regular than the others and is raised to a level above the original ground level. The excavator, Oscar Broneer, believes that the stone packing marks the outer, northeastern edge of an artificial spectator embankment in the earliest phase of the stadium.[55] Since the stone packing was relatively modest, the embankment would have been low in elevation with a small degree of slope. To the southeast of the stadium, the Cyclopean wall would have formed the outer limit of the spectator area, on naturally rising ground, if one existed at all. Although a racecourse floor must have existed, none was discovered. The date of the Archaic phase of the Earlier Stadium is difficult to determine since no datable pottery was recovered from the stone fill of the spectator area. Broneer has suggested 584-580 B.C. as an upper limit of construction since the 49th Olympiad is the historical date for the reorganization of the Isthmian Games as Panhellenic.[56]

The Classical phase of the Earlier Stadium at Isthmia has more architectural evidence than that of the Archaic phase. The artificial spectator embankment on the northeast side of the stadium was a much more formal and substantial construction. Two parallel walls of ashlar masonry run parallel to the stadium floor for a distance of approximately 78 meters; the inner wall is 23 m. to the east of the eastern water channel of the *dromos*. This wall which is preserved in places to three courses and to a total height of 1.50 m. begins 9.10 m. east of the Long Altar and was most likely continuous. At the northwest end of the inner wall, an angle of about 63 degrees is formed with a short spur wall which extends only 4.45 m. to the south. The spur wall acted as a part of the retaining system for the northwest end of the embankment in the Classical phase of the Earlier Stadium. The outer wall was built parallel to the first at a distance of over three meters and is stepped in as many as four preserved courses. The slope of the artificial eastern embankment has been restored as 8 degrees

52. The Sanctuary of Poseidon at Isthmia was excavated between 1952-1960 by the University of Chicago in cooperation with the American School of Classical Studies at Athens. The following discussion is based largely on the final publication of the stadia by Oscar Broneer, *Isthmia*, Volume II, *Topography and Architecture*, Princeton (1973): 46-66.

53. Two recent articles bear on the subject of the Isthmian stadia; Elizabeth R. Gebhard, "The Early Stadium at Isthmia and The Founding of the Isthmian Games," *Proceedings of an International Symposium on The Olympic Games (1988)*, William Coulson and Helmut Kyrieleis, eds., Athens

(1992): 73-79; Elizabeth R. Gebhard and Frederick P. Hemans, "University of Chicago Excavations at Isthmia, 1989: I," *Hesperia* 61 (1992): 1-77.

54. Whereas Broneer, *Isthmia* II (1973): 46-47, distinguishes essentially two phases of the Earlier Stadium, Gebhard and Hemans (above note 53) p. 57, note 131, suggest four phases.

55. Broneer, *Isthmia* II (1973): 46-47.

56. Broneer, *Isthmia* II (1973): 65.

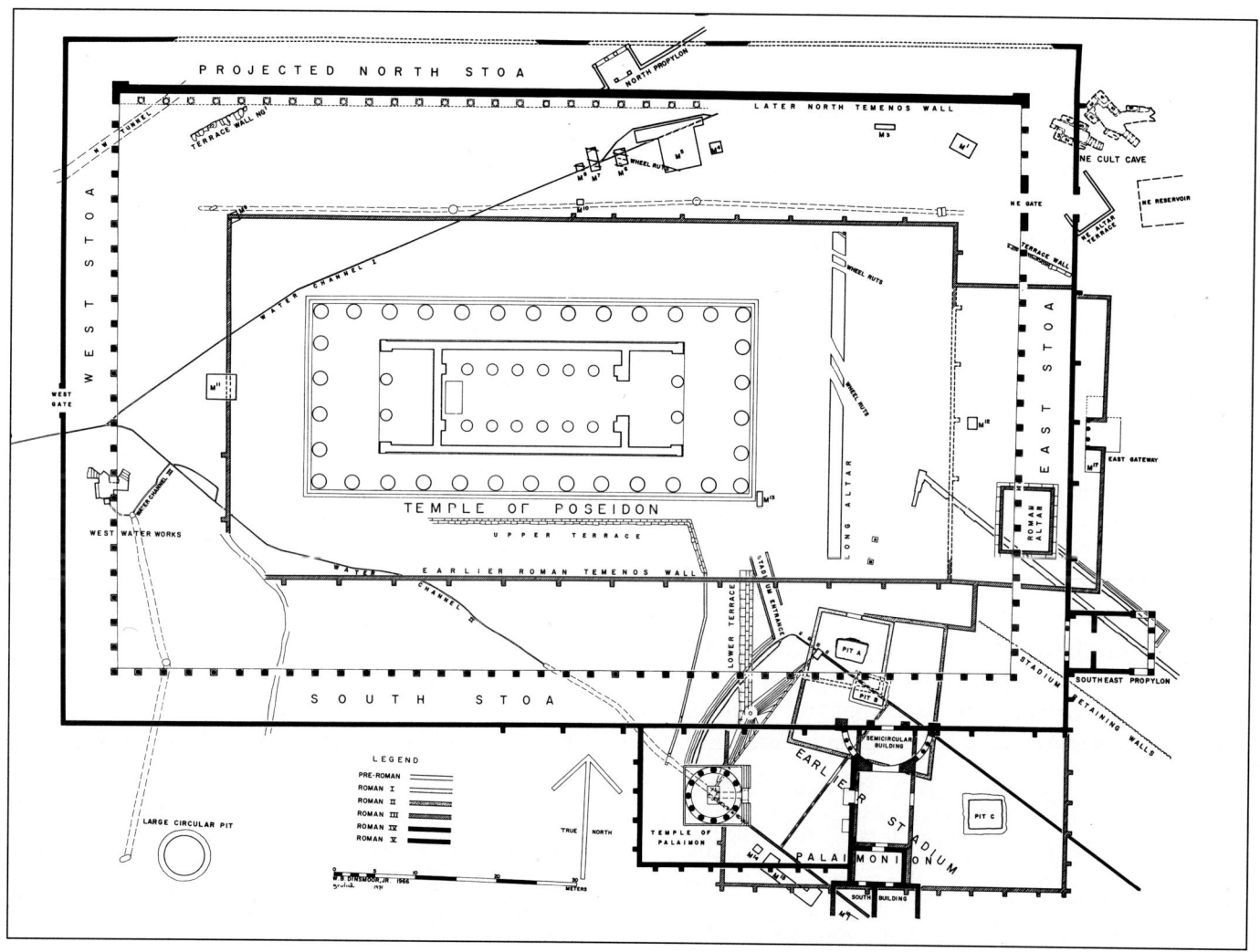

15. Isthmia, Sanctuary of Poseidon. Temple of Poseidon and the Earlier Stadium. Plan courtesy of the University of Chicago, Isthmia Excavations.

and 22 minutes. No seats have been found on the artificial embankment nor are there any seats associated with it. Three bases were found on the southwest side of the Earlier Stadium and are labeled M[14] M[15] and M[16] (Fig. 15). It has been suggested that these three monuments may have been bases for *proedria* or perhaps were the reserved seats of the judges.[57] Broneer has suggested that the curved and irregularly terraced steps at the northwestern end of the Earlier Stadium may have been used for spectators to stand on. I have estimated the spectator capacity of the artificial embankment and the terraced steps as approximately 4,000.

Two starting lines, and possibly the robbing trench for a third, were found at the northwest end of the Earlier Stadium (Fig. 16). The most ancient of the starting lines, dating to the fifth century B.C. is a triangular stone pavement (called by Broneer the *balbides* sill) which is only partially preserved.[58] It is a stone sill 20.42 m. in length, with cuttings to receive upright wooden posts, a starter's pit in the axis, and a series of sixteen grooves radiating outward from the pit (Fig. 17a). The stones of the triangular pavement vary in size and shape but all are approximately 0.15 m. high. The horizontal stone sill which abuts onto the triangular pavement, provides cuttings for eighteen vertical posts. These post holes are found at average intervals of 1.05

m. providing 16 starting positions with an additional blank central lane 1.35 m. wide (Fig. 17b). There are no toe grooves for the athlete's feet in association with the stone triangular pavement or the horizontal sill. The vertical posts had two functions, to provide visible starting position dividers and to provide starting gates for 16 athletes. A discussion of the working of the starting gates follows in Chapter 4, pp. 81-83.

During the excavation of the area, a layer of white clay, 0.05 meters thick, covered the triangular pavement and the horizontal sill and extended 20 meters to the southeast, to the point where the racecourse was disturbed by later construction. Within the layer of white clay was found what appeared to be a trench left after the robbing of a line of blocks, 0.70 meters wide and only 0.93 meters to the southeast of the horizontal sill. Broneer suggests that this trench may represent a line of starting line blocks which replaced the triangular stone pavement.

Also found within the layer of white clay was a second starting line, 10.93 meters to the southeast of the horizontal sill and parallel to it.[59] This starting line has a single groove cut in sections on the top surface of the blocks as well as post holes lined with lead.[60] The groove is approximately 0.07 meters wide, 0.04 meters deep and the post holes

---

57. Monument M[14] consists of a single block, 1.00 x 1.00 X 0.40 m. which is oriented parallel to the stadium water channel and approximately 1.00 meter to the south. Monument M[15] is a *pi*-shaped monument, with a total length of ca. 8 meters, located immediately to the southeast of Monument M[14]. Monument M[16] which appears to be a rectangular monument has been exposed for a length of 3 meters and shows a width of ca. 2.30 m.

58. The Greek word *balbis* meant a starting gate for the racecourse, or more generally a starting point. See Broneer, *Isthmia II* (1973) Appendix II, "ΒΑΛΒΙΣ, ΎΣΠΛΗΞ, ΚΑΜΠΤΗΡ," p. 137-142. Broneer, *Isthmia* II (1973): 65, suggests that the date of the *balbides* sill may be 470-460 B.C. at the time of the rebuilding of the temple, "or not much later." Gebhard and Hemans (above note 53) p. 59, note 132 put the construction of the triangular stone pavement in the second half of the fifth century on the basis of excavated deposits from 1989, to be published.

59. A possible reason for moving the starting line 10.93 m. further to the southeast was to provide greater overrun space for finishing athletes, if the starting line was also used as a finishing line.

60. Starting lines with single grooves cut into their top surfaces are also found at Delphi in the Gymnasium *paradromis*, as well as outside of the south retaining wall of the stadium at Delphi. For the Gymnasium at Delphi, see Jean Jannoray, "Le Gymnase," *FdD*, Topographie et Architecture, Part III, Vol. II (1953) and for the Stadium at Delphi, see Pierre Aupert, "Le Stade," *FdD*, Topographie et Architecture, Vol. II, part 7, Paris 1979. A starting line block with a single groove cut into its top surface has been found in the Xenon at Nemea. For the Nemea block, see David Gilman Romano, "An Early Stadium at Nemea," *Hesperia* 46 (1977): 27-31. A second single groove starting line block was found at Nemea in the area of the bath building and is reported in Stella G. Miller, "Excavations at Nemea, 1982," *Hesperia* 52 (1983): 93-95. A number of starting line blocks with single grooves have been identified at Labraunda in the stadium. For the Stadium at Labraunda see Paavo Roos, "Labraunda och de antika stadionanlaggningarna," *Svenska Forskningsinstitutet i Istanbul Meddelanden* 2 (1977): 23-29. Four starting line blocks with single grooves have been noticed which are built into the bastion wall of Ayasoluk which encloses the Isa Bey Mosque at Ephesos. For these blocks see Paavo Roos, "wiederverwendete Startblöcke vom Stadion in Ephesos," *ÖJh* 52 (1978-): 109-113.

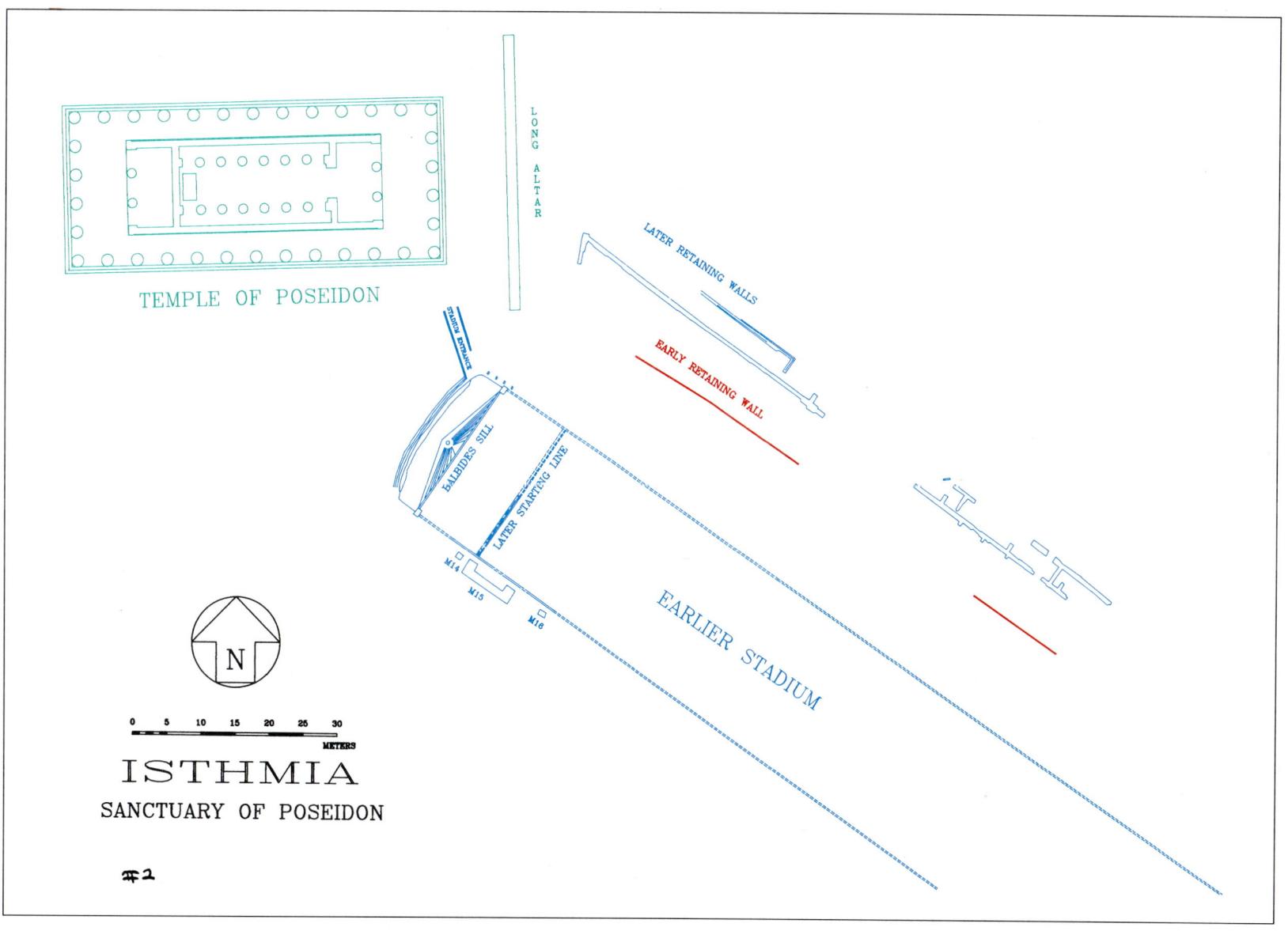

TEMPLE OF POSEIDON

LONG ALTAR

LATER RETAINING WALLS

EARLY RETAINING WALL

STADIUM ENTRANCE

BALBIDES SILL

LATER STARTING LINE

EARLIER STADIUM

M14

M15

M16

N

0 5 10 15 20 25 30
METERS

ISTHMIA

SANCTUARY OF POSEIDON

16. Isthmia, Sanctuary of Poseidon. Earlier Stadium, with retaining walls of embankments. Drawing by Gus Fahey after the plan by the University of Chicago.

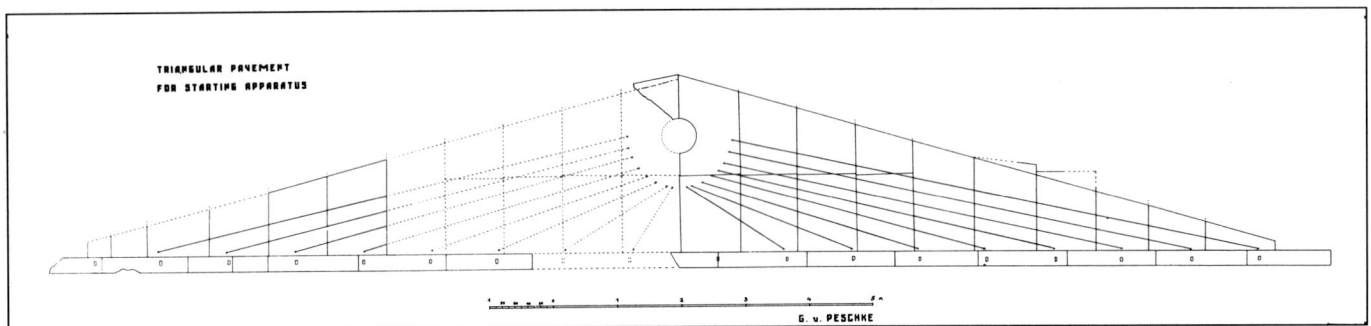

17a. Isthmia, Earlier Stadium, *balbides* sill, restored. Plan courtesy of the University of Chicago, Isthmia Excavations.

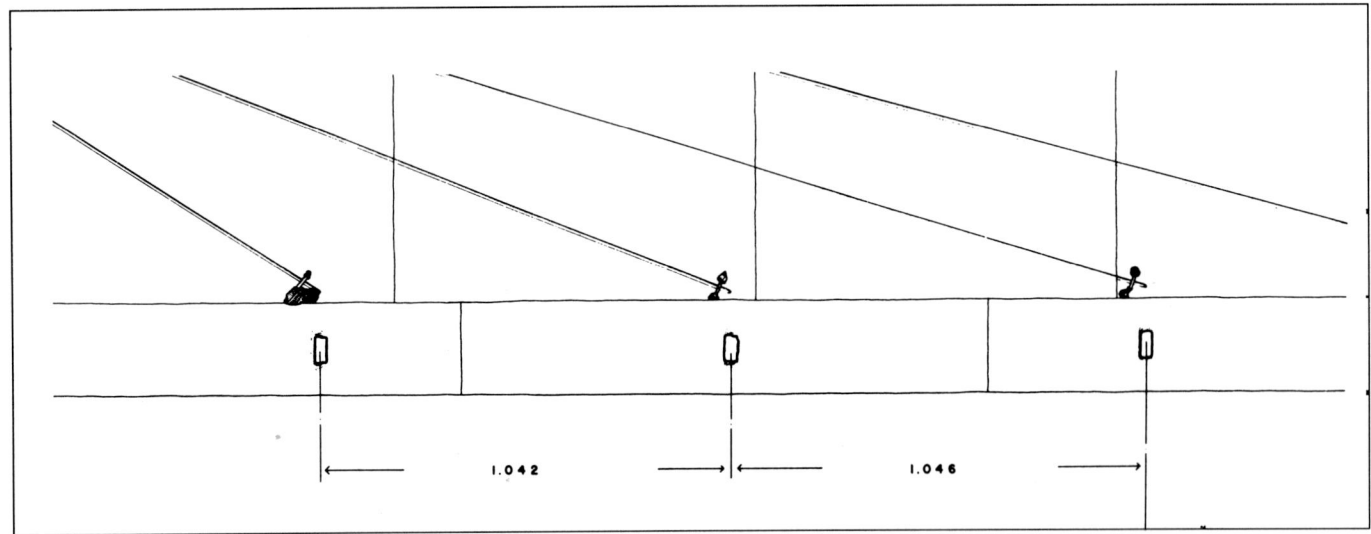

17b. Isthmia, Earlier Stadium, *balbides* sill (detail). Plan courtesy of the University of Chicago, Isthmia Excavations.

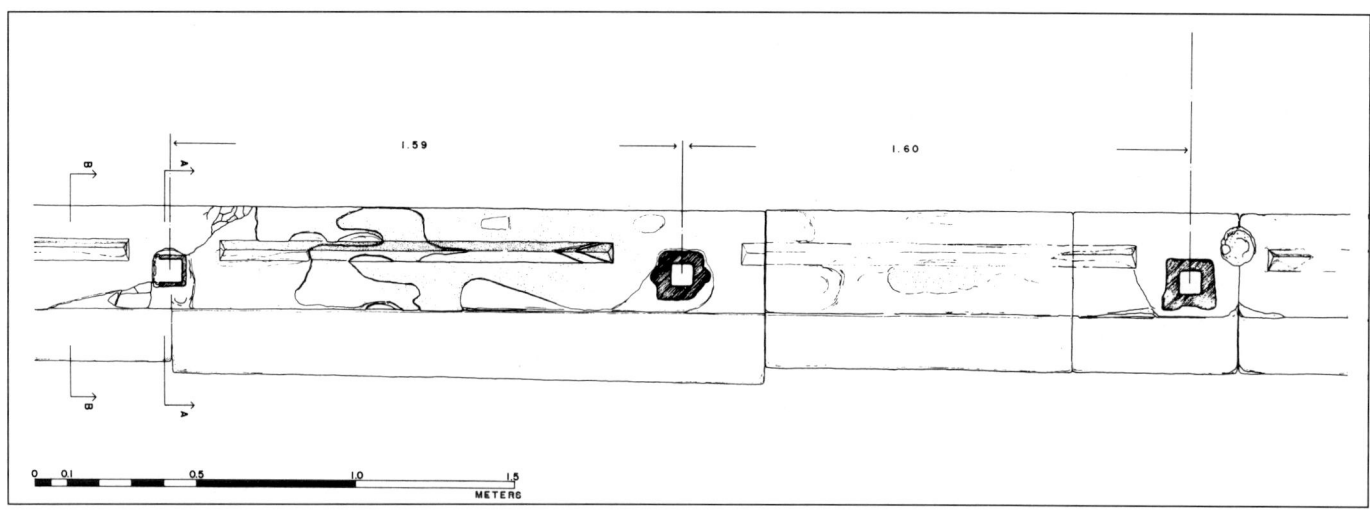

18a. Isthmia, Earlier Stadium, single groove starting line. Drawing courtesy of the University of Chicago, Isthmia Excavations.

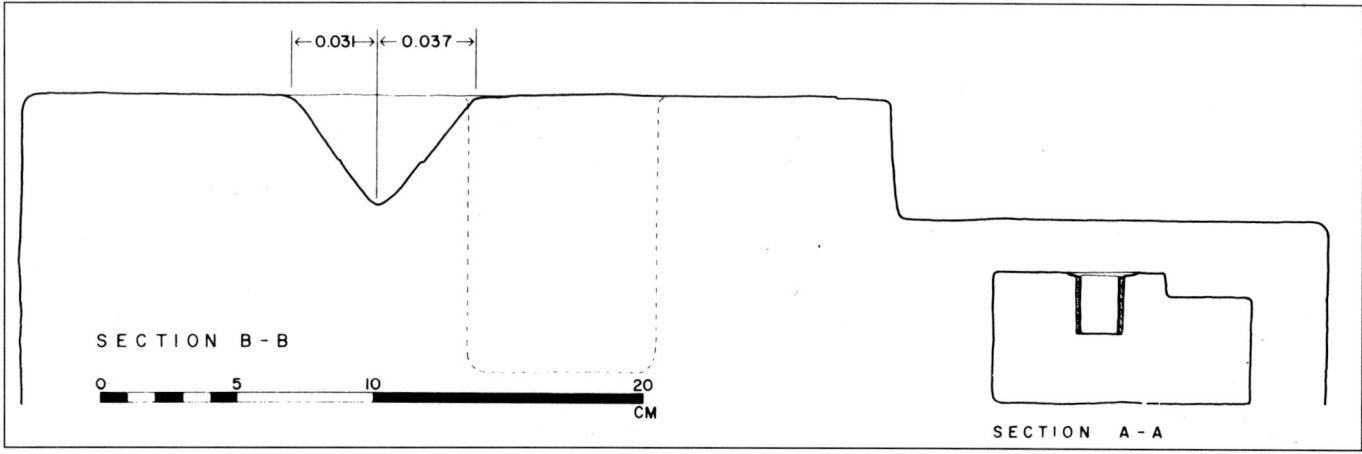

18b. Isthmia, Earlier Stadium, single groove starting line (section). Drawing courtesy of the University of Chicago, Isthmia Excavations.

are approximately 0.08 meters square (Fig. 18a,b). Although only six meters of the starting line are undisturbed, the line is preserved for a total length of 11.60 meters. The blocks of the line are of poros stone covered with a thin layer of cement. Broneer has restored twelve lanes, each 1.595 meters wide, with an additional central blank lane 2.23 meters wide. Although the date of this phase of the starting line is not certain, Broneer has suggested the early fourth century B.C.

A water channel originating from a westerly direction fed water as far as a reservoir near the northwest end of the stadium. A conduit led water into the stadium at the southern extreme of its closed western end. Water circulated in two directions from this point: around the curved end of the stadium in two water channels to the northern long side, and along the southern long side. Two large stone water basins are found at either end of the triangular pavement. An oval basin 0.87 x 0.66 m. and 0.31 m. deep is at the southwest end of the starting line, and a rectangular basin 1.02 x 0.73 x 0.40 m. deep is at the northeastern end of the line.

The formal entrance to the curved northwest end of the stadium was by means of an entrance ramp which began about 5.00 m. south of the southeast corner of the Temple of Poseidon, and led down to the northern corner of the curved end of the stadium. The ramp, flanked by two low walls, was approximately 11 m. in length and 1.90 m. in width. The incline of the ramp was 1 in 11 or 5 degrees, 12 minutes.

The Later Stadium at Isthmia, located in a natural hollow about 240 m. southeast of the Temple of Poseidon, is known only from a series of trenches and tunnels dug in the 1960s and from its vis-

ible contours (Fig. 19). Pausanias (II,1,7) mentions the stadium in his visit to Isthmia.

The Later Stadium which was probably constructed in the Hellenistic period, possibly during the time of Alexander, and then refurbished in the Roman period, has two slightly concave long sides and a semicircular closed end towards the southwest. The racecourse floor is bordered on three sides by the natural embankments of the hollow and is open at the northeast end. Starting lines have been found at both the northern and southern ends of the stadium, 181.20 m. apart.[61] More is known about the northern starting line which has been restored as ca. 26.60 m. in length. It is made of poros blocks which have two parallel grooves on their top surfaces. The grooves which are measured center to center as 0.13 m. apart, are interrupted for the cutting of post holes, 0.07 m. square, lined with lead casings. The northern starting line shows evidence of two phases of development. In the first phase there are sixteen post holes and seventeen starting positions. In the second phase, a large block 1.26 x 0.55 m. was joined to the racecourse side of the line opposite the second starting position from the northeast. The block is a part of the *husplex* mechanism, similar to those found in the stadia of Epidauros, Nemea, Olympia and Corinth.[62] One may restore a second *husplex* block at the other end of the starting line, which provided 12 starting positions between with a blank central lane. Only a small section of the southern starting line was excavated although rectangular statue bases were found at both ends of the southern starting line.

The racecourse is limited to the east and west by a narrow stone water channel, as well as a series of basins, cut of poros blocks

---

61. This measurement is made between the front edge of the post holes at the northern line and the middle of the southern line.

62. See below Chapter 5, pp. 85-86.

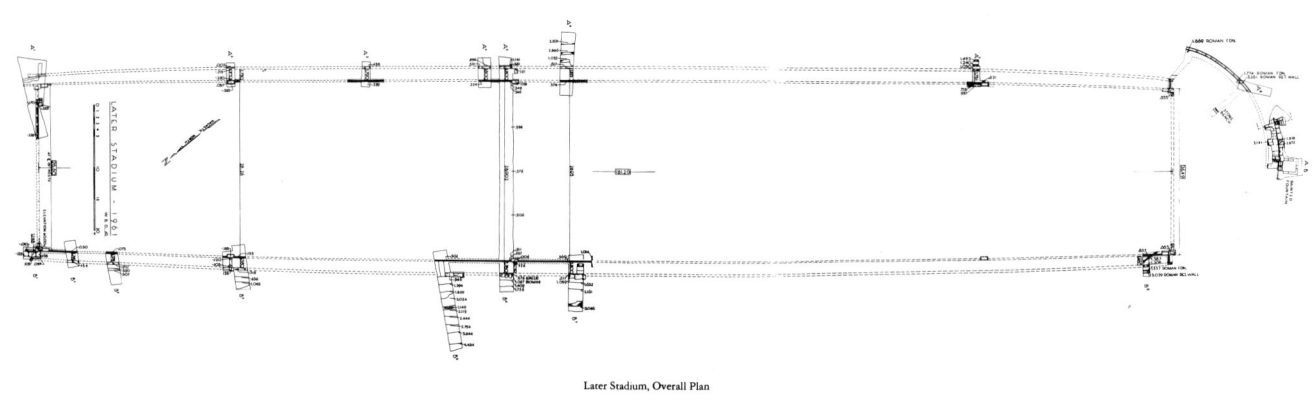

Later Stadium, Overall Plan

19. Isthmia, Later Stadium. Overall plan. Drawing courtesy of the University of Chicago, Isthmia Excavations.

and lined with stucco. The water channels were fed with water originating from a fountain house and reservoir at the southern extremity of the stadium, the source of which is not known. Between the narrow stone water channels and the beginning of the spectator embankment is a broad trench, approximately 1.295 m. wide which must have been used as a drainage channel. The surface of the racecourse floor was found to be composed of a hard packed clay surface which slopes down from south to north at a gradient of approximately 0.5 percent.

The spectator facilities were investigated in a number of trial trenches and found to be composed on the west side of a series of relatively flat, stepped terraces cut into the clay bedrock.[63] I have estimated the spectator accommodation of the stadium to be approximately 21,000. Broneer has associated a series of blocks set into the sloping side of the embankment together with blocks on the racecourse floor as blocks used by the judges to sit on.

63. Sections of some of these terraces are illustrated by Broneer, *Isthmia* II (1973) plate 60.

HALIEIS

Another example of an early stadium is from the Sanctuary of Apollo outside the ancient city of Halieis, located near the tip of the Southern Argolid (Fig. 3). Excavation of the city (above ground) and the sanctuary (under water) was undertaken in the 1960s and 1970s by the University of Pennsylvania and Indiana University under the direction of Michael H. Jameson.[64] Today, considerable areas of Halieis exist both above sea level and below and the Sanctuary of Apollo is outside of the city walls, 600 meters to the northeast of the city and totally submerged beneath the Bay of Porto Cheli.[65] The sanctuary was not a Panhellenic sanctuary; the games celebrated in the stadium were local games and the stadium is the only excavated example from mainland Greece of a stadium in a small, local sanctuary.

The stadium is located south of the Temple of Apollo, a second temple, an altar and associated buildings (Fig. 20). The axis of the stadium is approximately north-south, agreeing generally with the axes of the temples and altar. The stadium is composed of a rectangular *dromos* and spectator facilities. The *dromos* is limited at the south and the north ends by starting lines characterized by two parallel grooves, 166.50 meters apart,[66] and along the entire east side by a stone curb. Stone foundations for artificial spectator embankments flank the *dromos* in the northern one-third of its length on both the east and west sides.

On the east side of the racecourse, at the northern end, are two roughly rectangular areas which border the *dromos*, labeled eastern embankments 1 and 2. Each rectangular area is defined by a line of stones, the lengths of which vary but the widths of which are fairly constant at 0.50-0.60 m. The stone foundations are preserved to only one course in height; the northern embankment measures 24.60 x 9.25 m. and the southern embankment measures 26.9 x 9.25 m. The two rectangular areas fall in elevation from east to west following the slope of the ground level. The stone foundations are likely to have demarcated an area that had earlier been used for watching the contests.

The *temenos* wall of the sanctuary, leading from the gate, meets the northern wall of eastern embankment 1 near its northwest corner and appears to have originally delimited entry into the spectator embankment from the north. Access to this embankment was by means of a passage in its northwest corner. Separating the two eastern embankments is a passage approximately 8.8 m. x 2.80 m. wide. It is most likely that the stones delimited low embankments on which the spectators would sit or stand. Within eastern embankment 2 is a line of stones parallel to the *dromos* with a total length of 17.35 m. joining a second line of stones 1.95 m. long. These stones may have supported a bench or platform for the judges.

On the west side of the *dromos* is the western embankment; there is preserved a double row of relatively large blocks running almost parallel to the curb of the racecourse and approximately 4.5

---

64. See Michael H. Jameson, "The Excavation of a Drowned Greek Temple," *Scientific American,* October (1974): 111-119; Romano, "Stadia" (1981): 30-52 and Michael H. Jameson, "The Submerged Sanctuary of Apollo at Halieis in the Argolid of Greece," *National Geographic Society Research Reports,* volume 14, Washington, D.C. (1982): 363-367. The final results of the Halieis Excavations will be published by Indiana University Press.

65. For a discussion of the planning of the city of Halieis, see Thomas D. Boyd and Michael H. Jameson, "Urban and Rural Land Division in Ancient Greece," *Hesperia* 50 (1981): 327-342.

66. The measurement was made from the forward groove of each starting line.

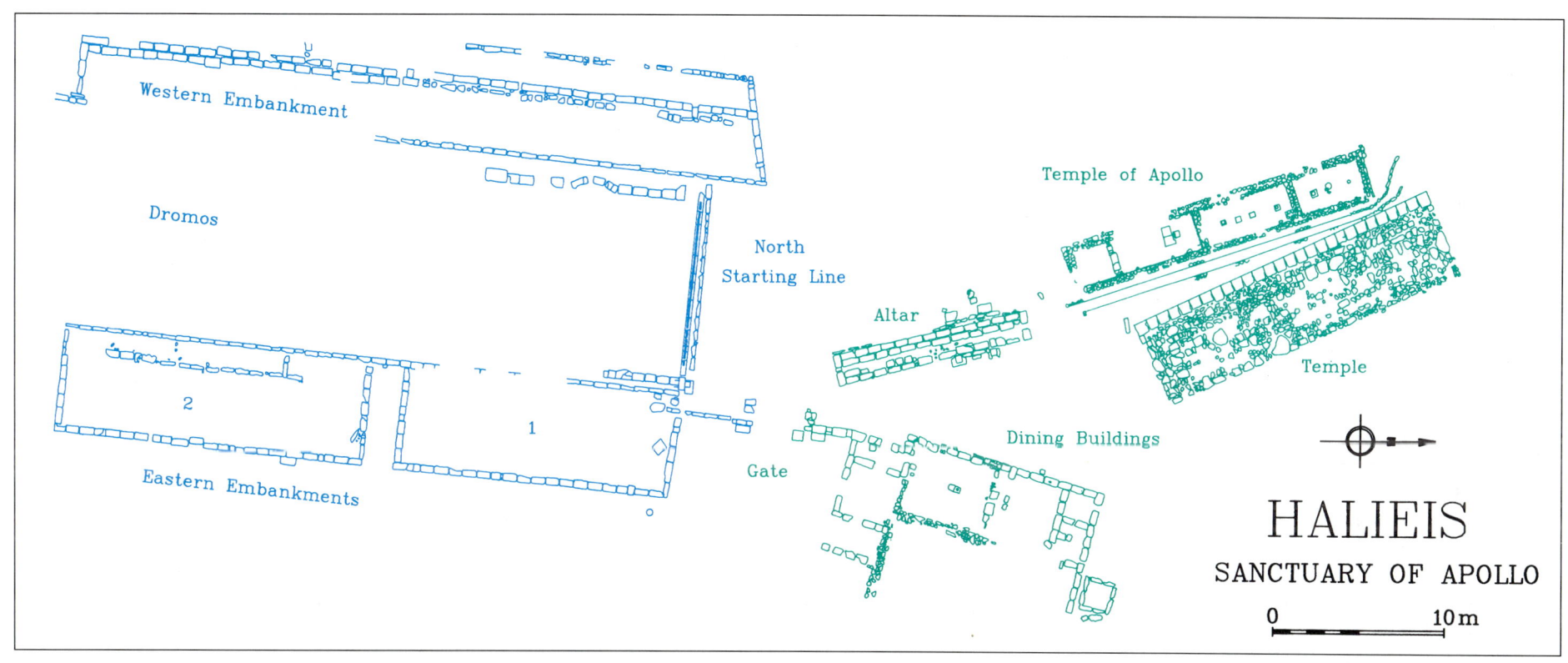

20. Halieis. Sanctuary of Apollo. Drawing by Mimi Woods after the plan by F.A. Cooper and T.D. Boyd, 1973.

m. to the west. Many of the blocks are large with lengths up to 2.40 m. and widths relatively constant at 0.50 to 0.60 m. Approximately 2.5 m. further to the west there begins a second line of blocks parallel to the double row. They are preserved for a total length of 25 m although they may be restored with a total length of approximately 58 m. Cross walls join the racecourse curbing with the double line of blocks in two places. Clearly the double thickness wall was meant to support something very heavy and it is likely that one or more courses of equally large blocks once rested on top of these to provide support for a back retaining wall for an earth embankment, artificially constructed, that would have sloped up from the level of the stone curbing of the *dromos* floor. Although the original slope of these embankments at Halieis is not known, it is likely that the slope would have been low. There is no evidence for stone foundations for seats. I have estimated the total spectator capacity at approximately 1,500.

Two starting lines limit the *dromos* to the north and south; the length of the northern starting line measures 15.50 m. and the southern starting line 16.42 m. The stone starting line blocks have two parallel grooves which are cut into the top surfaces of the 12 blocks of the northern starting line and the easternmost 5 of the 16 blocks of the southern starting line. The parallel grooves are interrupted by square post holes providing for 7 lanes, each approximately 2 m. in width, in both the north and south starting lines. The two parallel grooves are each approximately 0.03 m. wide and 0.14 m. apart, measured center to center. There is no evidence for a *husplex* at Halieis.

Although there is no conclusive ceramic evidence for the specific construction date of the stadium, black-glazed pottery frag-ments found in the racecourse floor suggest that the stadium may be no earlier than the sixth century B.C. There is no evidence for hydraulic facilities of any kind in the Halieis stadium. No water channels or water basins were found, nor was there any trace of drainage facilities.

The best parallels for the type of artificial embankments known at Halieis are from Olympia and Isthmia, described above. The construction of the embankments at Halieis appears to be a part of the same building project that includes the gate, dining buildings and possibly the altar, all generally dated to the Classical period.

It is possible, therefore, that the *dromos* of the stadium at Halieis was laid out at the same time that one of the two temples was built (each temple appears to have been built as a separate project, but both were built by ca. 600 B.C.) and that a nearby level stretch of ground was employed for a racecourse for the athletic contests held in connection with the cult. In the beginning spectators stood on the gradually sloping ground to the east of the *dromos* and later, probably in the fifth century B.C., the artificial embankment was added to the west side of the racecourse as a more formal spectator area. The two parallel groove starting lines (commonly found in Hellenistic and Roman stadia) were likely introduced later still, as replacements for earlier ones.

Archaeological Conclusions

All of these early, sixth and fifth century B.C., stadia are similar in that they are found within the limits of the sanctuary and they

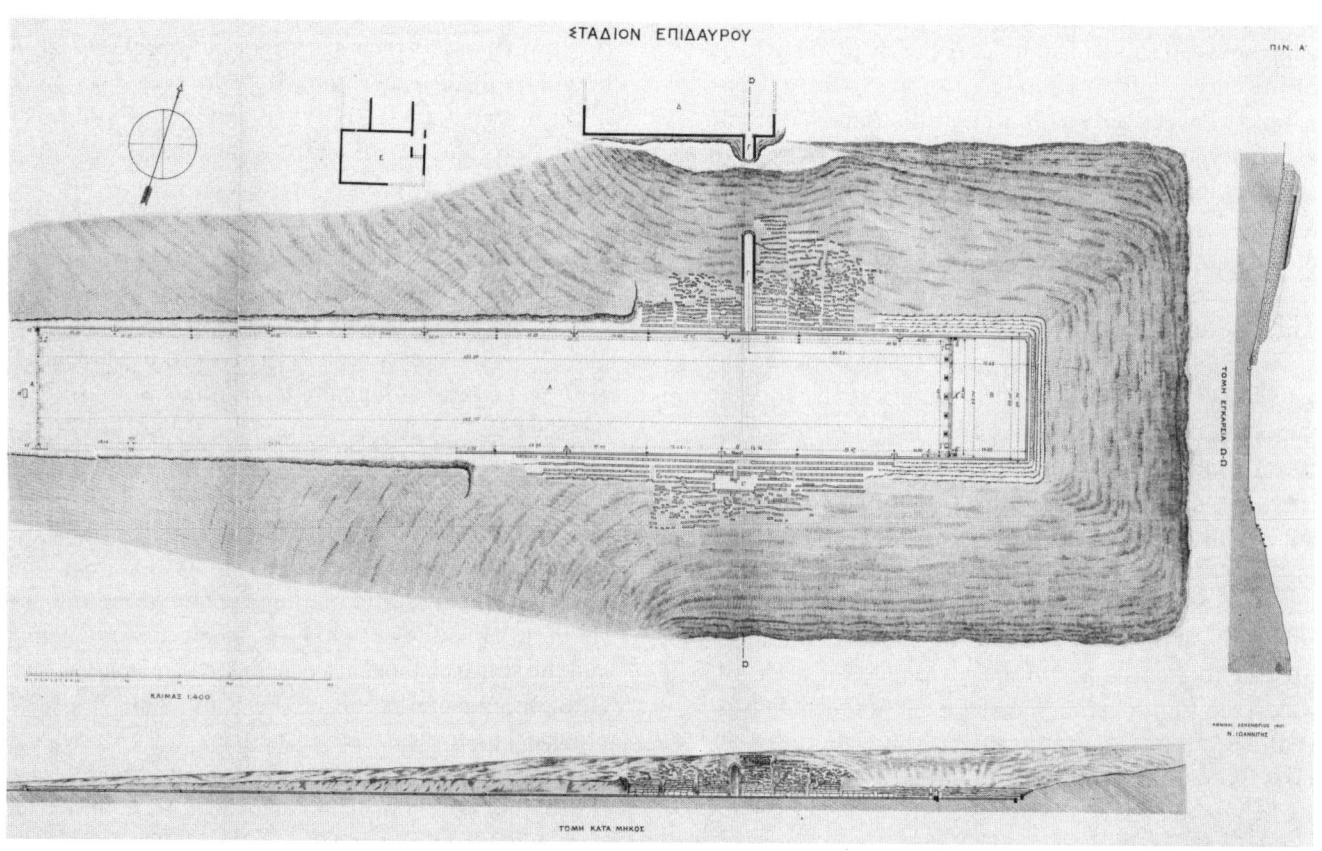

21. Epidauros, Sanctuary of Asklepius, Stadium. Plan courtesy of the Archaeological Society of Athens.

are located near the principal temple and altar of the site.[67] This close physical relationship between cult and athletics is clearly fundamental in the location of the early stadium although it seems to have been less important as time went on and stadia were often removed from the heart of the sanctuary. As a result, the *dromos* became less important, relatively speaking, having been moved to an area better suited to the spectator.

The earliest (sixth century) racecourse floors have not been found at either Olympia or Isthmia and, therefore, the slope of the artificial embankments cannot be accurately calculated. The slope in both cases, however, would have been very gentle. The slope of the artificial embankments is known from the fifth-century stadia at both Olympia and Isthmia. At Olympia the slope of the northern embankment is 9 degrees or 1/6, and the slope of the southern embankment is 4 degrees or 1/12. At Isthmia the slope of the northern embankment is 8 degrees and 22 minutes. The low slope of these embankments thus suggests a more comfortable surface on which to stand rather than sit. In addition, the curved and irregularly terraced steps at the north western end of the stadium have been suggested as standing room for spectators. There exists additional evidence for areas of standing room within the fourth century stadia from both Epidauros and Nemea. At the stadium at Epidauros in the Sanctuary of Asklepius, which is likely to have been built in the first half of the fourth century B.C., there are a series of five shallow terraces cut into the lower slopes of the eastern, northern and southern embankments,

some with small flat stones resting on the top surfaces of the terraces; the lowermost terrace has larger cut blocks which form the eastern boundary of the racecourse at floor level (Figs. 21-23).[68] The height of these terraces is 0.20-0.30 m. and their depth is 0.50-0.70 m. I have estimated that approximately 750 spectators could have stood on these terraces.

These shallow terraces may be easily contrasted with the ashlar stone seat blocks that were installed in the northern and southern embankments during the Hellenistic period in the same stadium (Fig. 22), where the blocks are 0.35-0.41 m. in height and the depth of the block plus the access area to the rear of the block is 0.79-1.06 m. suggesting a comfortable area to sit on as well as a lateral passageway between seat blocks.

In the late fourth century B.C. stadium at Nemea there exists as many as six artificial terraces cut into the clay bedrock at the southern end of the stadium.[69] The height and width of these terraces vary considerably (0.20-0.40 m. deep and 0.15-0.35 m. high) suggesting an area more easily stood on than sat on. Along the western side of the stadium, outside the line of the western stone water channel blocks and the storm channel are two and in some places three rows of reused stone blocks which follow the convex curve of the water channel blocks. The top surface of the first row of blocks is approximately 0.10 m. above the top level of the stone water channel and the second row of blocks is approximately 0.20-0.25 m. above the top of the first row of blocks suggesting its use for standing room. Only a few blocks

---

67. There are other examples of early stadia being situated close to the temple and principal altar of the site, e.g., Tegea and Nemea. See Romano, "Stadia" (1981): 179-184; 186-187, and Stella G. Miller, "Excavations at Nemea, 1982," *Hesperia* 52 (1983): 93-95.

68. See P. Kavvadias, "Τὸ Στάδιον, Ἀνασκαφαὶ ἐν Ἐπιδαύρωι, *Praktika* 1900-1902, Athens, p. 80 and Romano, "Stadia" (1981): 19-22.

69. Stephen G. Miller, "Excavations at Nemea, 1977," *Hesperia* 47 (1978): 84-88, fig. 7 and Romano, "Stadia" (1981): 88-89.

22. Epidauros, Stadium. Photograph looking east. Photo courtesy of DAI, Athens, neg. nr. EPI 76.

23. Epidauros, Stadium. Detail photograph of east embankment. Photo courtesy of the Archaeological Society of Athens.

of the third row remain.

This evidence further suggests areas of standing room in Greek stadia at least into the fourth century B.C.[70] Of course the evidence from Olympia suggests primarily standing room in the largest of all of the excavated Peloponnesian stadia through the Roman period.

The origin and early architectural development of the Greek *stadion* may have followed an evolution similar to that of the ancient Greek theater where the orchestra, from the Greek word ὀρχεῖσθαι, literally the "dancing place," is the center of attention as the spectators sat watching. The theater, from θεάσθαι, "to see" was really a structure built for the spectators, although the word "theater" came to mean the entire area for both spectators and performers.[71] The earliest form of the *stadion* was likely to have been as a *dromos* bordered by a simple and parallel embankment of earth, either natural or artificial. Later there would have developed more substantial embankments for greater spectator space and eventually the embankments bordered the *dromos* on more than one side and literally "wrapped around" the *dromos*.[72] Sometimes this created a rectilinear embankment, as is the case from the stadium at Epidauros, and at times a curvilinear structure with one rounded end, as at Isthmia.

The athletes needed only the *dromos* or running area on which to compete, whereas it was the spectators who needed the *stadion* in order to watch the contests. Likewise the word *stadion* came to mean the total of the areas for the spectators as well as for the athletes. Thus the comparison is suggested: theater/orchestra = *stadion/dromos*. In both cases the name of the whole structure may have derived not from the area where the activity took place, but from the name of the area built for the spectators. Facilities for spectators are a *sine qua non* for the ancient *stadion*. Without spectator facilities, the structure is a *dromos*. During the Hellenistic and Roman periods it was common to find stadia with slightly convex curved sides as well as curved ends as at the Later Stadium at Isthmia, the late fourth century B.C. Stadium at Nemea and the Roman Panathenaic Stadium at Athens. The curves in the design were likely related to providing better spectator viewing for the athletic events, both at the ends of the stadium and along the sides.[73]

Since one of the principle meanings of the word *stadion* is as a linear distance equal to 600 feet, it is useful to consider briefly here the length of the *stadion* in the sixth and fifth centuries. The length of the *dromos* of the Olympia III Stadium as 192.28 m. between starting lines is known from direct measurement in the modern day, although the starting lines are likely to be from a later (Roman) renovation. We do not know from archaeological evidence the lengths of the Olympia I or the Olympia II Stadia, although they are generally assumed to have been the same length.[74] Similarly, we know the length of the *dromos* of the Later Stadium at Isthmia from the mea-

70. Similar terraces have been found in the cavea of the Greek theater at Elis. See Veronica Mitsopoulos-Leon and Erwin Pochmarski, "Elfter Vorlaufiger über die Grabungen in Elis," *ÖJh*, 51 (1976-1977): cols. 181-222.

71. Elizabeth R. Gebhard, "The Form of the Orchestra in the Early Greek Theater," *Hesperia* 43 (1974): 428-440.

72. I thank Homer A. Thompson for discussing this idea with me and for suggesting the parallel architectural development of the theater and the stadium.

73. The purpose of the convex long sides of Hellenistic and Roman stadia is clear from the design of the Hellenistic stadium at Priene. Only one of the long sides of the stadium, the north side, has a spectator embankment and its design is slightly concave to the racecourse floor. The south long side is straight. See Theodor Wiegand and Hans Schrader, *Priene*, Berlin, 1904, plate 19. This principle and design of spectator accommodations may have been a contributing factor towards the shape and form of the Roman amphitheater. For another discussion of the subject, see Katherine Welch, "Roman Amphitheatres Revived," Introduction (above p. 1) note 1 pp. 274-277.

74. See above note 41.

sured distance between the starting lines, but not the length of the Earlier Stadium.[75] At Halieis we know from direct measurement the length of the *dromos* of the stadium between starting lines is 166.50 m., although the starting lines found *in situ* may be of a later renovation of the sanctuary. Of course, stadia lengths are known from other sites, but can be documented only from later periods[76]

The Archaic and Classical stadia from Olympia, Isthmia and Halieis were all similar since they included a *dromos*, embankments of earth as spectator accommodation and were located in the heart of the sanctuary near the principal temple and altar. The Archaic and Classical racecourse, *dromos*, could also be found without artificial embankments, and possibly removed from temple and altar, as will be seen in Corinth.

75. Oscar Broneer, *Isthmia*, Volume I, *Temple of Poseidon*, Princeton, 1971, Appendix I: The Foot Measure pp. 174-181, has suggested that the Earlier Stadium at Isthmia was originally measured as the same length as the stadium at Olympia, 192.28 m, and then reduced in length to 181.20 m. See below p. 92, n. 24 .

76. A full discussion of foot lengths in connection wiith stadia and starting lines found in the Peloponnesos is found in Romano, "Stadia" (1981): 250-267.

Chapter 2
THE ARCHAIC *DROMOS* IN CORINTH

A unique curved starting line of a *dromos* from ca. 500 B.C., has been excavated in Ancient Corinth; its unusual design has importance not only for an understanding of athletic activity in Archaic and Classical Corinth but also for the history and evolution of Greek and Roman athletics and athletic architecture for centuries to follow.[1] The starting line in Corinth, and the associated features of the *dromos*, constitute the earliest example from the Greek world that can be securely dated.[2]

During the Spring of 1980, the curved starting line was excavated in the area immediately to the west of the Julian Basilica in the area of the east forum (Fig. 24).[3] A small portion of the same starting line had been found in 1936-1937 when the neighboring, straight Hellenistic line was excavated although at that time, it was not realized that the cleared portion of the earlier line was a part of a curve.[4] The early starting line is associated with an Archaic and Classical racecourse that was situated in the broad and fairly flat area to the south of Temple Hill and the Sacred Spring, the Upper Lechaion Road Valley (Fig. 25).[5] The earliest floor of this racecourse can be dated no earlier than the sixth century B.C.[6] The racecourse extended approximately 165 meters in length to the southwest, to the other end of the valley, where it is limited by a north-south roadway with an open drain. Approximately 158 meters southwest of the starting line, along the southern side of the *dromos* is an uninscribed rectangular "horos" type stone stele *in situ*. The upper dimensions of the stone are 0.20 x 0.26 m. Approximately 0.85 m. to the south is a roughly rectangular block with a rectangular recess cut into its top surface with dimensions 0.86 x 0.59 m. (Fig. 27a,b). Both the "horos" type stone and the rectangular block are set into a chipped poros floor which rests on top of a harder chipped stone layer. The floor can be dated probably to the late sixth century B.C. The "horos" stone may have served as a distance marker on the south side of the racecourse, as a marker at the west end or as a marker for a small shrine. The entire length of the Archaic and Classical racecourse was crisscrossed by roadways and bordered by shrines which seemed to characterize this area of the city in the Archaic and Classical periods. Spectators could have watched the contests on the racecourse from rising ground to the south of the racecourse. In the modern day most of the length of the Archaic *dromos* is not visible, having been covered by later fill as

1. Although I refer to the early starting line at Corinth as Early Classical, ca. 500 B.C., the range of possible dates for the starting line are ca. 500-450 B.C. See below note 17. Other early *in situ* and datable starting lines include the first preserved starting line at Isthmia, the triangular stone *balbides* sill, that dates to after 470-460 B.C., (Broneer, *Isthmia,* II (1973): 65-66), above Chapter 1, pp. 23-32; the starting line in the Athenian Agora, dating to the mid-fifth century B.C., T. Leslie Shear, Jr. "The Athenian Agora: Excavations of 1973-1974," Hesperia 44 (1975): 362-365.

2. There are two other *dromoi* from mainland Greece, portions of which are excavated, one from the Athenian Agora and one near the agora in Argos. For a plan of the *dromos* in Athens see John M. Camp, *The Athenian Agora, Excavations in the Heart of Classical Athens,* New York (1986): 45-46, fig. 66. For Argos see Marcel Piérart and Jean-Paul Thalmann, "Agora: zone du Portique," BCH 102 (1978): 777-783 and Romano, "Stadia" (1981): 188.

3. For a full and detailed discussion of the excavation of the Early Classical Starting Line and neighboring areas by Charles K. Williams, II, Director, Corinth Excavations of the American School of Classical Studies at Athens in Spring 1980 see Williams (1980): 1-44.

4. Charles H. Morgan, II, "Excavations at Corinth, 1936-1937," AJA 41 (1937): 549-551, pls. XVI, XVII.

5. The word racecourse is commonly used at Corinth for each of the two successive "tracks" that have been discovered in this area of the Greek city. There is a general lack of permanent or temporary spectator facilities excavated to date at Corinth in connection with the two successive tracks, thus the nomenclature "racecourse" rather than "stadion." See below, note 18. It is likely, however, that spectators would have taken advantage of the gradual slope to the south of the *dromos* as a natural spectator slope, similar to the situation at Olympia in the seventh and early sixth centuries B.C., before the construction of the artificial southern embankment. See discussion above, Chapter 1, pp. 17-22.

6. Williams (1980): 2.

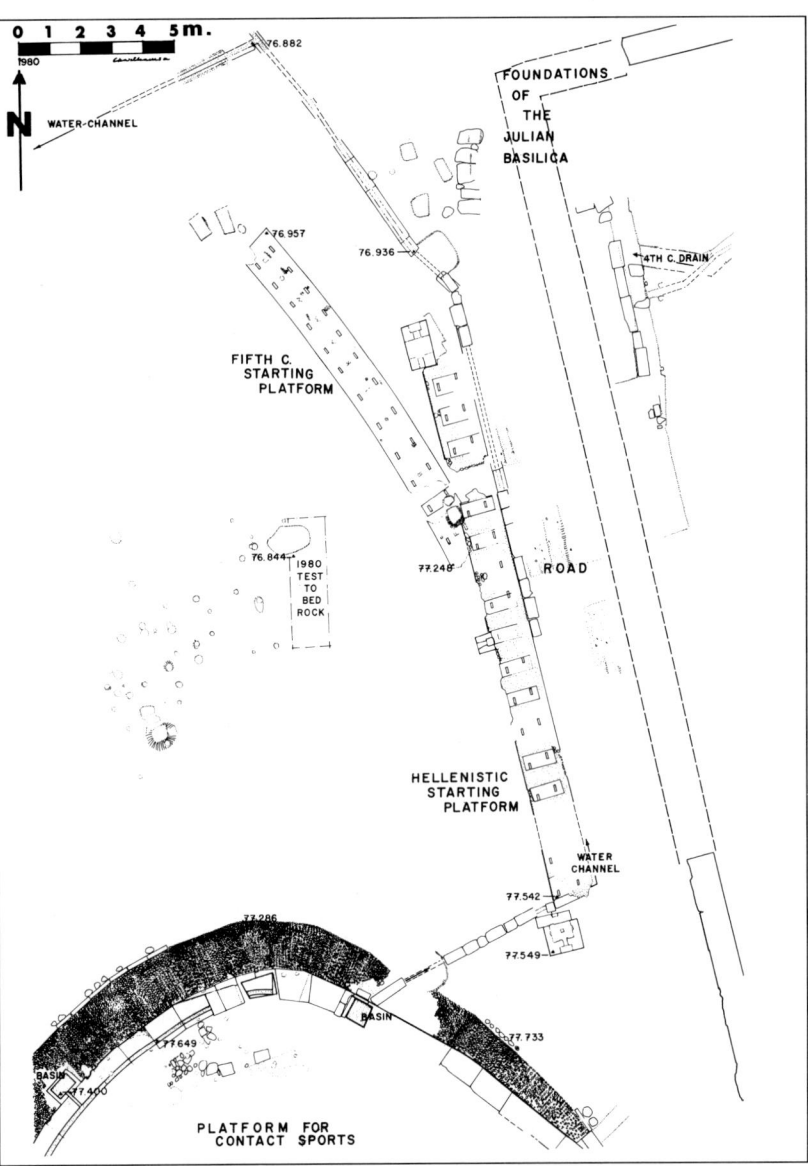

24. Corinth, Actual-state plan of Early Classical starting line, labeled fifth century starting platform, and Hellenistic starting line in the northeast corner of the forum in Corinth. Courtesy of the American School of Classical Studies at Athens, Corinth Excavations.

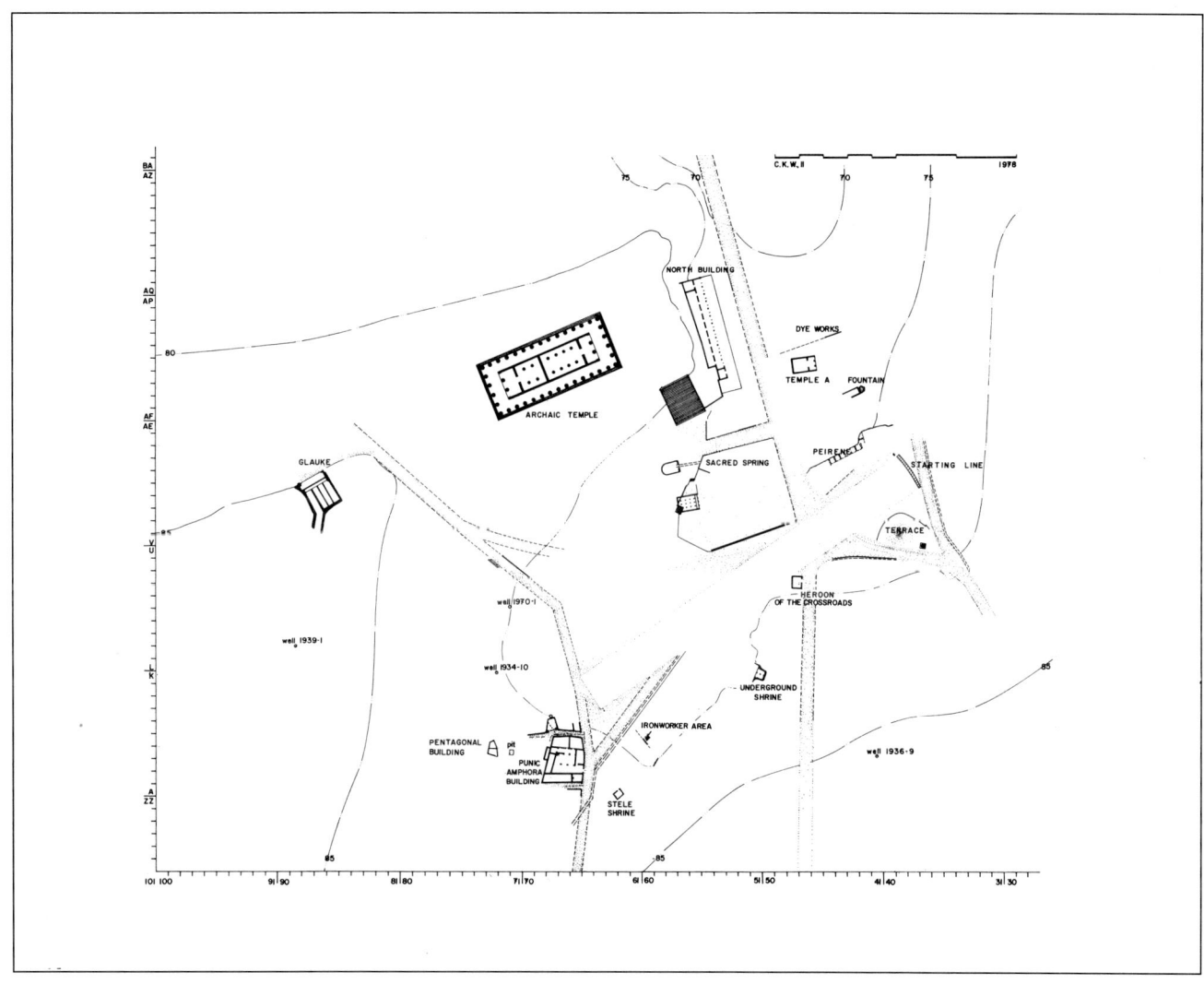

25. Plan of central Corinth, 450-425 B.C. Courtesy of the American School of Classical Studies at Athens, Corinth Excavations.

well as by the floor of the Roman forum.

Associated with the Early Classical starting line, immediately to its south is a curved terrace built out of the natural slope of the hillside (Figs. 24, 25, 26).[7] The terrace is bordered to the north by a low wall of ashlar blocks and a cobbled sidewalk at its base. A water channel and water basins were added to the terrace, probably as a part of alterations of the Hellenistic period. The terrace was first identified as a viewing stand for spectators and judges at the athletic contests.[8] More recently, Williams has suggested that the terrace was used for athletic competitions, such as the *pankration*, wrestling and boxing and that spectators could have viewed the contests from wooden stands, ἴκρια, constructed on the racecourse floor as well as from the southside of the terrace.[9]

Physical Characteristics of the Early Classical Starting Line

The early starting line is between 1.25 and 1.30 m. wide and is constructed of a pebbly, white lime cement. The length of the excavated starting line is over 12 m; the length was probably originally between 16 and 17 m (Fig. 28). The top surface of the starting line is smooth and was painted a dark blue-black in a fresco technique. The starting line is constructed of rectangular poros blocks of limestone with a thin cement laid between the stone blocks, a similar technique to the adjacent Hellenistic starting line.[10] Individual grooves, for the

toes of the front left foot and the rear right foot, were cut into each of the poros blocks after which the top surface of the starting line was plastered with the added blue-black pigment. A faint line is incised on the perimeter of the curved starting platform, on all sides, outside of which the cement has been painted white, so that the blue-black pavement would have been clearly visible against the white border. Red letters, approximately 0.05-0.077 m. long, indicating lane numbers, were painted on the finished surface of the line between each pair of front and rear toe grooves (Figs. 29, 30). The starting positions for the athletes were numbered from *alpha* at the south to *pi* at the north end. The five southernmost positions were destroyed by the construction of the Hellenistic starting line above. The first preserved starting position to the south shows a *vau*, followed by a *zeta*. The next position shows no letter, but then follow *heta, theta, iota, kappa, lambda, mu, nu, omicron,* and *pi.* The letter *xi* does not appear in the series after *nu* (Figures 29, 30).[11] The letters were meant to be read from the east side of the starting line, looking towards the west end of the racecourse.

Evidence for thirteen starting positions are visible from the 12.20 m. excavated portion of the starting line. Each of these positions originally consisted of two individual toe grooves, for the front (left) foot and for the rear (right) foot. Each groove has a vertical back wall and a beveled front wall.[12] The rectangular toe grooves vary somewhat in size, 0.215 to 0.23 m. in length, and they are consis-

7. Williams (1980): 15-21.

8. Charles Morgan, II, "Excavations at Corinth, 1936-1937," *AJA* 41 (1937): 550-551.

9. Williams (1980): 15-21.

10. Such is the construction of the poros blocks of the Hellenistic starting line that is immediately adjacent to the fifth century B.C. line, the construction of which is more easily visible. See below Chapter 5, pp. 85-95.

11. Williams notes that there are indications of repainting of some of the letters on the starting line, suggesting that the line was refurbished after a certain degree of use. See Williams (1980): 7-8.

12. This was a common feature in starting line design. The vertical back wall provided a surface for the toes of the foot to push off against and the beveled front face of the groove guaranteed that the athlete would not catch his toes on the groove as he left his position.

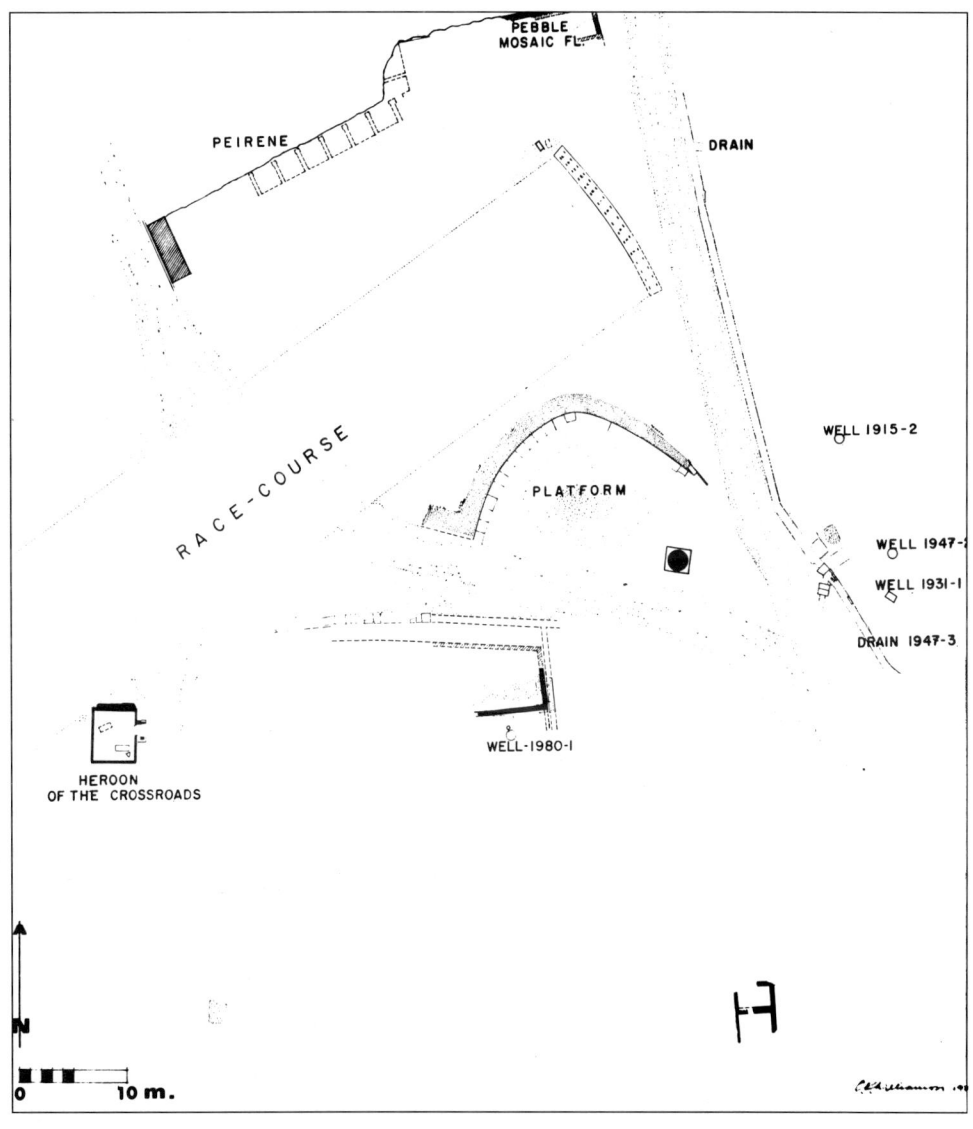

26. Corinth, detail of eastern end of archaic *dromos* and neighboring terrace, here labeled "platform."
Courtesy of the American School of Classical Studies at Athens, Corinth Excavations.

tently 0.075 m. wide. The depth of the grooves varies between 0.045 to 0.069 m. The distance between the front and rear groove at each starting position varies considerably from 0.595 to 0.87 meters.[13]

The widely spaced toe grooves of each starting position are virtually unique in Greek athletics. In addition the starting line did not include any post holes that would provide lane dividers or turning posts or attachments that might suggest a mechanical starting device. There appears to have been no need for these elements since the individual toe grooves restricted the location of the starting athlete at the starting line and also guaranteed that the athlete be fairly stable at the start as a result of the required wide stance.[14] It would be difficult for an athlete to lean too far forward and loose his balance.[15]

Evidence exists for a second starting line of the late fifth century or early fourth century B.C. being built over the Early Classical starting line.[16] The second starting line was slightly higher than its

predecessor and was originally surfaced in blue-black and then resurfaced in white cement.

Racecourse Surface

The excavations of 1980 at Corinth exposed the successive running surfaces of the racecourse as a series of crushed poros levels. These are preserved up to, and flush with, the ca. 500 B.C. starting line.[17] It has been suggested that the compacted poros was spread seasonally as required by the necessities of the cult races. A number of post holes were excavated in the surface of the racecourse floor, 5-10 meters to the west of the curved starting line, near the central axis of the track (Fig. 24).[18] As mentioned above, there was discovered a similar crushed poros floor in the area near the southwestern end of the Archaic *dromos* which can also be dated to the late sixth century B.C.[19]

13. Starting lines with individual toe grooves of this particular type are not common. They have been found elsewhere only at Nemea in two separate blocks. See David Gilman Romano, "An Early Stadium at Nemea," *Hesperia* 46 (1977): 27-31 and Romano, "Stadia" (1981): 179-184. See also Stella G. Miller, "Excavations at Nemea, 1982," *Hesperia* 52 (1983): 93-95. Miller dates the block and its use to the Hellenistic period due to the appearance of, and the form of, a single *lambda* located on the top surface of the block. It is possible, however, that the block may originally have been used earlier than this.

14. The same variety of starting position is also visible in the straight Hellenistic starting line at Corinth (below Chapter 5, pp. 85-95) although the space between the front and rear toe grooves is more regular at ca. 0.56 m. Since the later line is likely to have been for a short race, the necessary conclusion must be that this particular variety of widely spaced foot position starting line is common to Corinth and that it was used between 500-146 B.C.

15. The only similar designs were known from nearby Nemea, above note 13. From a passage in Herodotus (8,59) there is a reference to problems of starting runners "jumping the gun" in the footrace. It is Adeimantos, son of Okytos the Corinthian admiral who, while discussing with Themistokles military strategy before the Battle of Salamis in 480 B.C., says "in the games,

those who start before the signal are flogged." At which Themistokles replies in defense, "but those that are left behind do not win the crown."

16. Williams (1980): 10-11. Williams suggests that the reason for the construction of the later raised starting line may have been related to the *dromos* being resurfaced as well as to make the toe groove intervals more equal.

17. See Williams (1980): 8-10 and note 8. Williams states that although the starting line is generally referred to in his excavation report as being of a fifth century date, the crushed poros racecourse floor and the starting line may have been constructed in the late sixth century B.C. There is no evidence, however, that they were necessarily laid down simultaneously. Although the curved starting line does not give the impression of being the first of its kind, Williams notes that if there was an earlier starting line on the same site, no trace of it remains.

18. Williams (1980): 6, fig. 4. These post holes may have been related to the temporary bleachers, *ikria*, that were set up to watch the field events that are suggested to have taken place on the circular terrace to the south of the Archaic *dromos*. Another possibility would be that the post holes are related to the footraces run on the *dromos*, and possibly could have been the locations of turning posts in the distance race over a period of years.

19. See above p. 43, and figs. 27 a, b.

27a. Corinth, photograph of stele and base at west end of Archaic *dromos*. View towards the north. Courtesy of the American School of Classical Studies at Athens, Corinth Excavations.

27b. View towards the south.

Reconstruction of the Design of the Early Classical Dromos and Starting Line

The steps in the design of the Corinth racecourse and starting line by an ancient Greek architect are fairly simple and can now be summarized. A scale drawing was likely made by the architect using a straight-edge and a compass in order to lay out the plan of the entire racecourse as well as the particular elements of the starting line and the "break line" before construction work was begun.[20]

By definition, a *stadion* is 600 feet long and therefore the space available in the Upper Lechaion Road Valley would have had to have been at least 600 feet long and of commensurate width. Although the exact length of the racecourse is not known from archaeological evidence, and the length of the Archaic Corinthian foot is not previously known it has been noted that the length of the racecourse should fall between 158 and 165 meters.[21] The widths of

20. The "break line" was the area on the racecourse where the runners who had started in individual lanes would be allowed to leave their lanes and run toward the inside lane of the *dromos*. See discussion below, pp. 53-62.

28. Corinth. Photograph of the curved Early Classical starting line looking towards the south. The straight Hellenistic starting line can be seen in the mid-ground of the photograph. Courtesy of the American School of Classical Studies at Athens, Corinth Excavations.

stadia vary, the only requirement being that it be wide enough for the number of runners and the other needs of the course.

It is possible to reconstruct the long sides of the Archaic *dromos* by drawing a hypothetical line between the northeast corner of the sixth century Hero Shrine that would have bordered the racecourse on its south side, and altered slightly for the racecourse, and the restored south end of the curved starting line (based on 17 starting positions).[22] The north long side of the racecourse can be drawn parallel to the south side, clearing the quadriga base to the north, and extending to join the north edge of the starting line (Fig. 31).[23]

The length and width of the *dromos* of *stadion* length, thus defined, would be divided into six equal parts, each 100 feet long (one plethron) (Fig. 32).[24] This could probably have been designed most

21. See Williams (1980): 9-10. I have considered elsewhere the question of the possible length of the Corinth foot and the resulting Corinth *stadion*. See Romano, "Stadia" (1981): 164-166. I suggested, based on the measurement of other Archaic and Classical buildings in Corinth and Isthmia, the length of 0.269 m. as the foot on which the *stadion* is measured as 600 feet or 161.46 m. Based on the measurements of the starting line and the associated information, discussed below, the foot length on which the racecourse is measured now appears to be closer to 0.275 meters, with a resulting *stadion* in the neighborhood of 165 meters. This distance is surprisingly close to the total length of the neighboring fourth century B.C. South Stoa which is 164.47 meters as measured on the stylobate. If this relationship is valid, this would qualify the South Stoa as an example of a *stadion* length stoa of which there are other examples in the Hellenistic world, e.g., a stoa at Miletos dating to 300 B.C. (H. Knackfuss, *Milet* 1.7 (1924): 21-47, 281-282). See Broneer, *South Stoa* (1954): 33. I have measured the length of the South Stoa, between the exterior faces of the antae, with the electronic distance meter in 1988 and recorded a distance of 164.374 m. Broneer's measurement between the same two points was 164.38 m. Additional information that may relate to the measure of the Corinthian foot, is given by Strabo (8,6,21) who mentions that Akrocorinth is a lofty mountain with a perpendicular height of three stadia and one-half. In the modern day, the elevation of the geodetic marker at the summit of the mountain is 573.86 meters above sea level which, when divided by three and one-half, equals 163.96 meters.

22. The average chord distance between each preserved pair of starting positions on the curved line is 0.951 m. See below pp. 57-63.

23. The quadriga base is known to have been in existence before the construction of the Hellenistic *dromos*, while the Archaic *dromos* was still in use. Its exact date of construction is not known. The long side of the Archaic *dromos* clears the restored southeast corner of the quadriga base by 0.08 m.

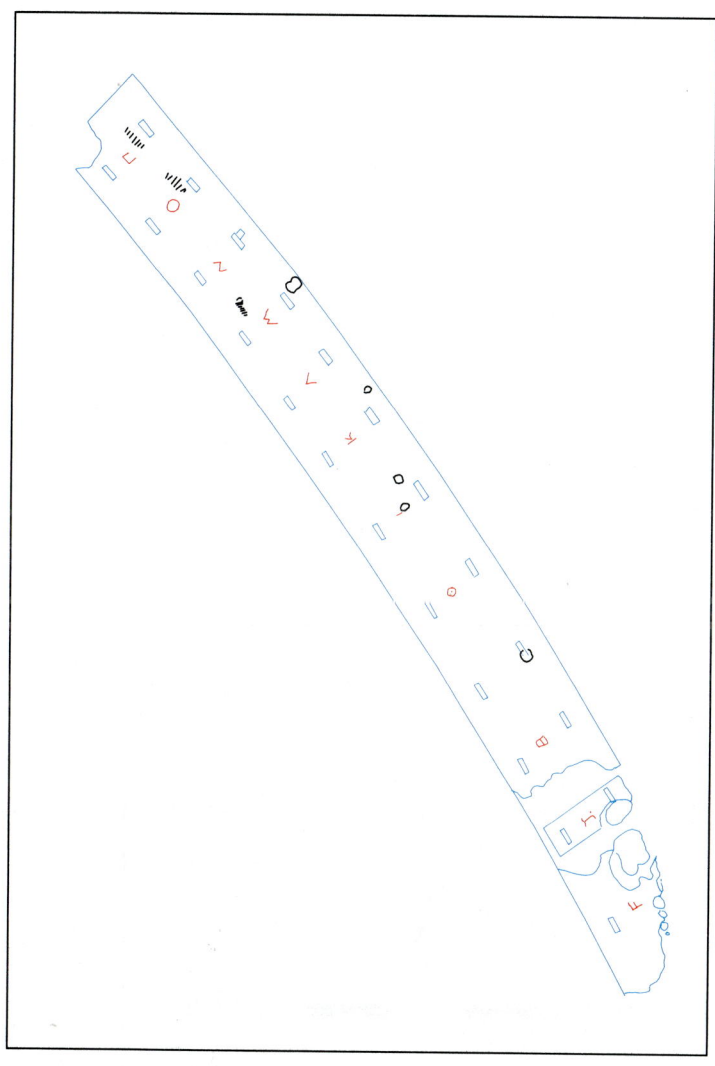

29. Drawing of the excavated portion of the curved Early Classical starting line including the red painted letters as lane numbers.

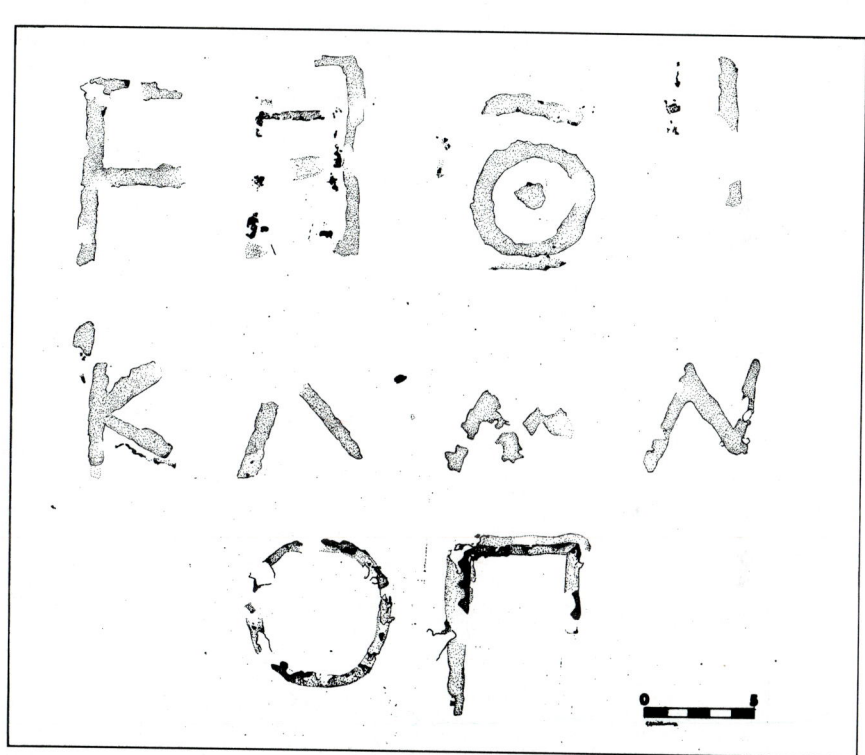

30. Drawing of painted letters on the surface of the Early Classical starting line. Courtesy of the American School of Classical Studies at Athens, Corinth Excavations.

52

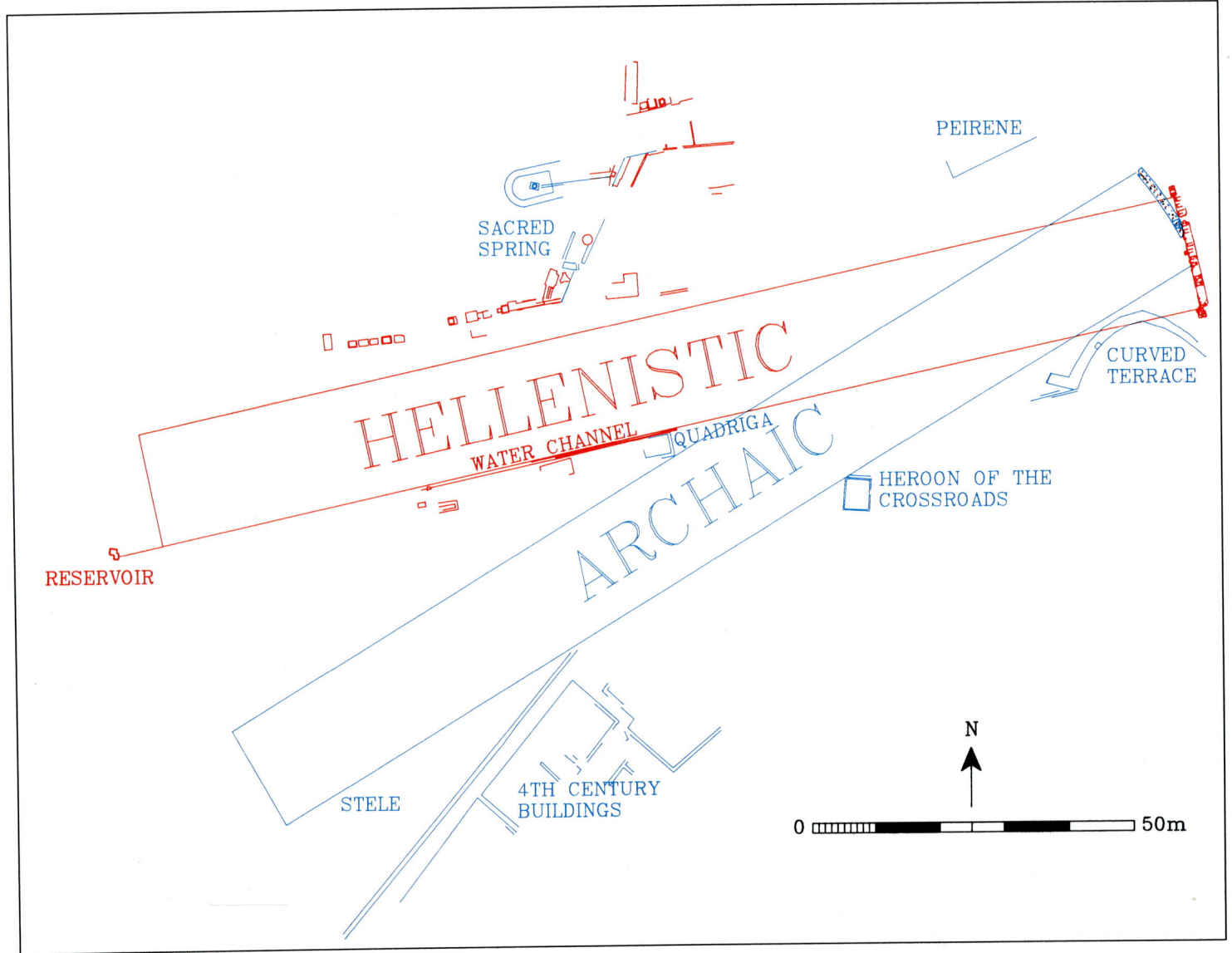

PEIRENE

SACRED
SPRING

HELLENISTIC

WATER CHANNEL QUADRIGA

ARCHAIC

CURVED
TERRACE

HEROON OF THE
CROSSROADS

RESERVOIR

N

STELE

4TH CENTURY
BUILDINGS

0 ⸺⸺⸺⸺ 50m

31. Drawing showing relative locations of the two successive starting lines, racecourses and neighboring monuments. The structures in blue are Archaic and Classical Greek, those in magenta are Hellenistic Greek.

easily from the center point of the racecourse, where a 100 foot, 200 foot and 300 foot arc could be drawn, thus creating the divisions of the racecourse into equal 100 foot sections. A circle would then be drawn of a 200 foot radius, using as the center a point on the 200 foot line, 10.26 meters from the north edge of the track and 6.30 meters from the south side (Fig. 33). This point was purposely selected by the ancient architect off of the central long axis of the track. An arc of this circle would create the curve for the starting line at the east end of the racecourse and the radius of this circle would be approximately 55 meters or 200 Corinthian feet.

The focus of the seventeen starting runners at the starting line would have been a point on the racecourse floor approximately one-third of the length of the racecourse.[25] The arc was not absolutely regular, whether measured from the front border of the starting line or from the toe grooves, which probably means that the starting line was laid out by hand. In order to study the nature of the arc, and to reconstruct the working methods of the ancient architect, a series of coordinates were taken in the field by means of the Electronic Total Station.[26] Twenty-one points were surveyed on the front incised line of the starting line from one end to the other. Two points at each end of the starting line were discarded because of irregularities. From the remaining seventeen points, 455 circles were created using all non-adjacent points. One center was totally out of range and was discarded. The centers of the circles created by these points were computed and the average of the 454 centers was plotted, the radius equaling 63.273 m. Thirty-two of the centers were then removed from the computation of the average because their north-south coordinate was extreme, and then the average of the centers was determined again, the radius equaling 55.274 m.[27] For this discussion I use the 55.274 m. radius.

The Greek architect purposely chose a curved starting line to provide a fair start and an equal distance to be run for each of the seventeen runners. Using a cord or line, the architect secured one end at this focal point on the racecourse floor; the other end he extended to the center of the starting line that he was to create. Between the

---

24. Some stadia have rectangular plinths as 100 foot markers set up along the long sides of the *dromos.* Such are visible, for instance, in the stadia at Nemea and Epidauros. One of the reasons for these *plethron* markers may have been related to the setting out of certain of the facilities of the *stadion* for different events, as at Corinth.

25. Williams (1980): 14-15, note 20. Williams concludes that the early starting line was surely designed primarily for runners at the start, as opposed to athletes in other events, since the curved design of the starting line and the numbered positions would make no sense otherwise. As mentioned above, note 13, the single toe grooves are not unique, however the curve of the starting line is.

26. The field equipment used in this study is the Leitz (Sokkia) Set-3 Electronic Total Station, which includes an electronic theodolite and an electronic distance meter, an SDR-22 handheld computer and the Lietz (Sokkia) MAP survey programs. The factory specifications of the equipment include an accuracy in distance to within 5 seconds of one minute of one degree and in angles to within 5 seconds of one minute of one degree. In a closed traverse in the area of the Upper Lechaion Road Valley on August 13, 1990, our precision was measured as 1:27200. The mathematic and geometric calculations have been achieved by the use of two computer programs, AutoCAD, an architectural drafting program, and DCA (Softdesk) Engineering Software, a civil engineering program. For a more detailed discussion of the methods employed in the Corinth Computer Project see David Gilman Romano and Benjamin C. Schoenbrun, "A Computerized Architectural and Topographical Survey of Ancient Corinth," *Journal of Field Archaeology,* 20 (1993): 177-190. I thank Benjamin Schoenbrun for mathematical assistance in the analysis of this curved starting line and for writing the computer program using Basic for the mathematical study. I also thank three graduate students in the Department of Mathematics at the University of Pennsylvania for their time and interest in discussing with me the geometric questions of the design: Andy Hicks, Sarah Nemeth and Richard Garfield.

27. By another method a 55.929 m. radius was calculated using the maximum interval density of the 454 centers and adjusting for minimum standard deviation.

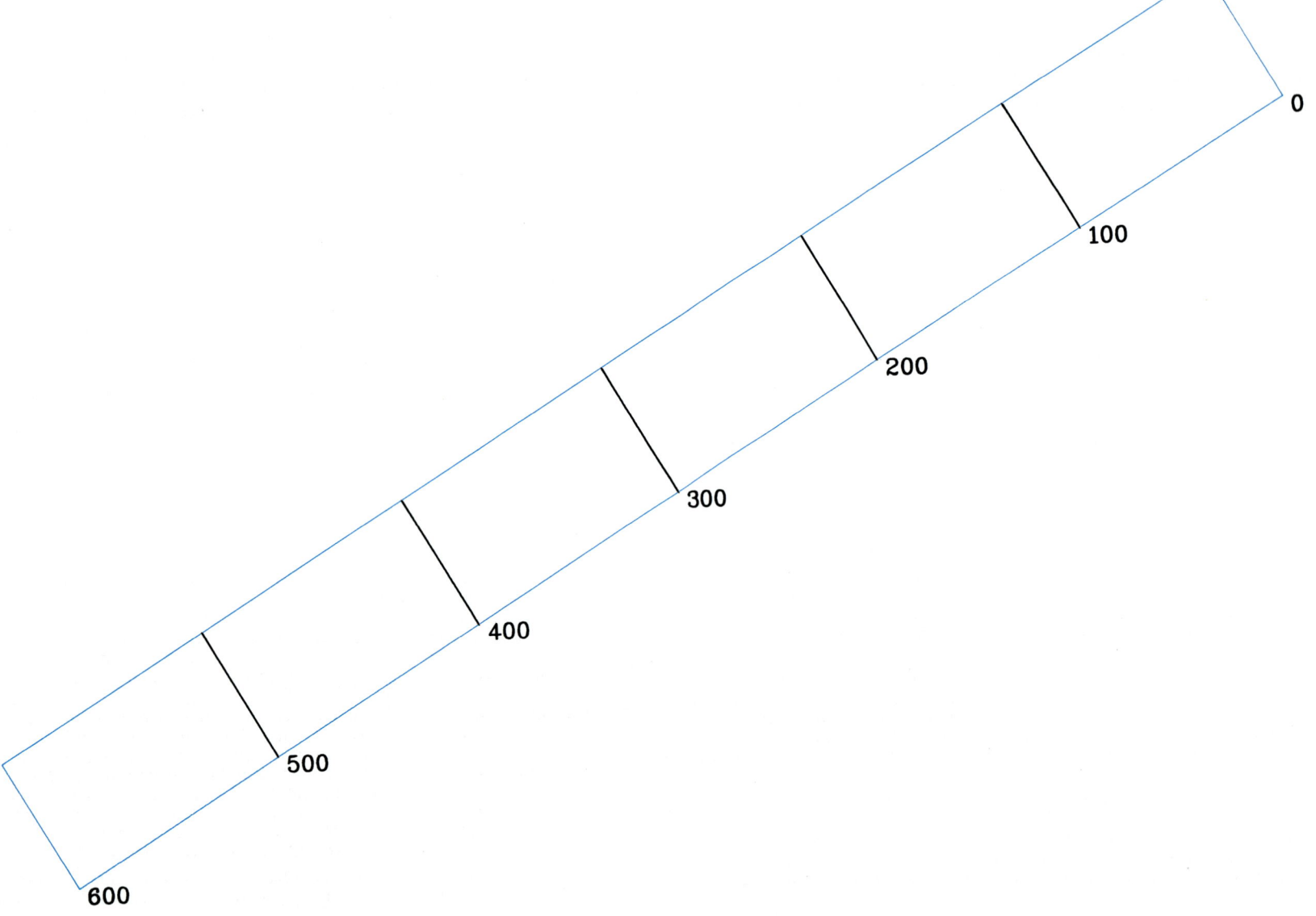

32. Stages of planning for the Early Classical starting line and Archaic *dromos:* 600 foot stadion is divided into six equal 100 foot lengths.

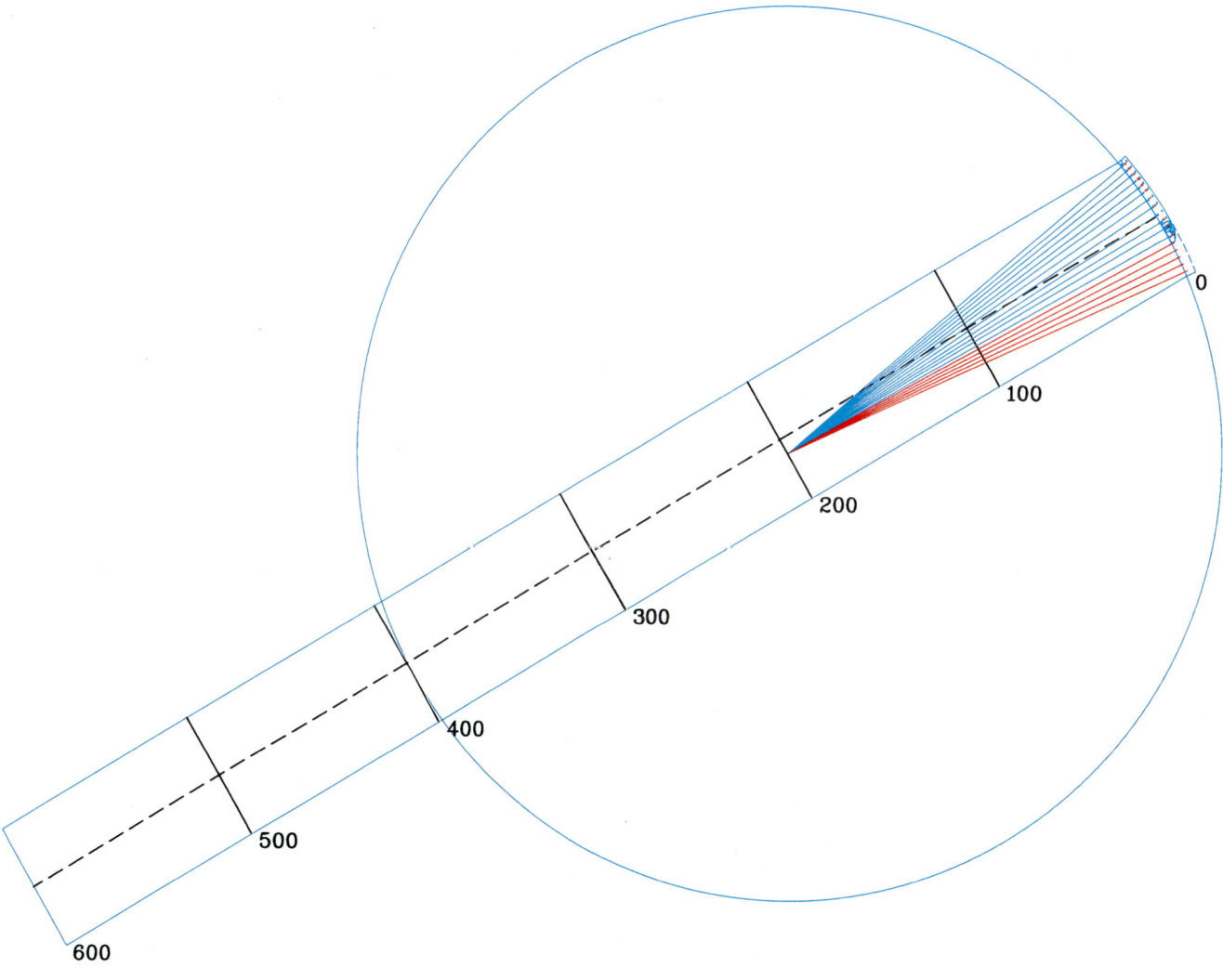

33. A circle with radius of 200 feet is centered on the 200 foot line, to the south of the midpoint of the track. Seventeen starting positions are set out on the arc of the starting line, each is equidistant from the focal point. Blue lines indicate the radius line to each of the excavated starting positions and red lines, to the unexcavated positions. The dashed line indicates the central axis of the racecourse floor.

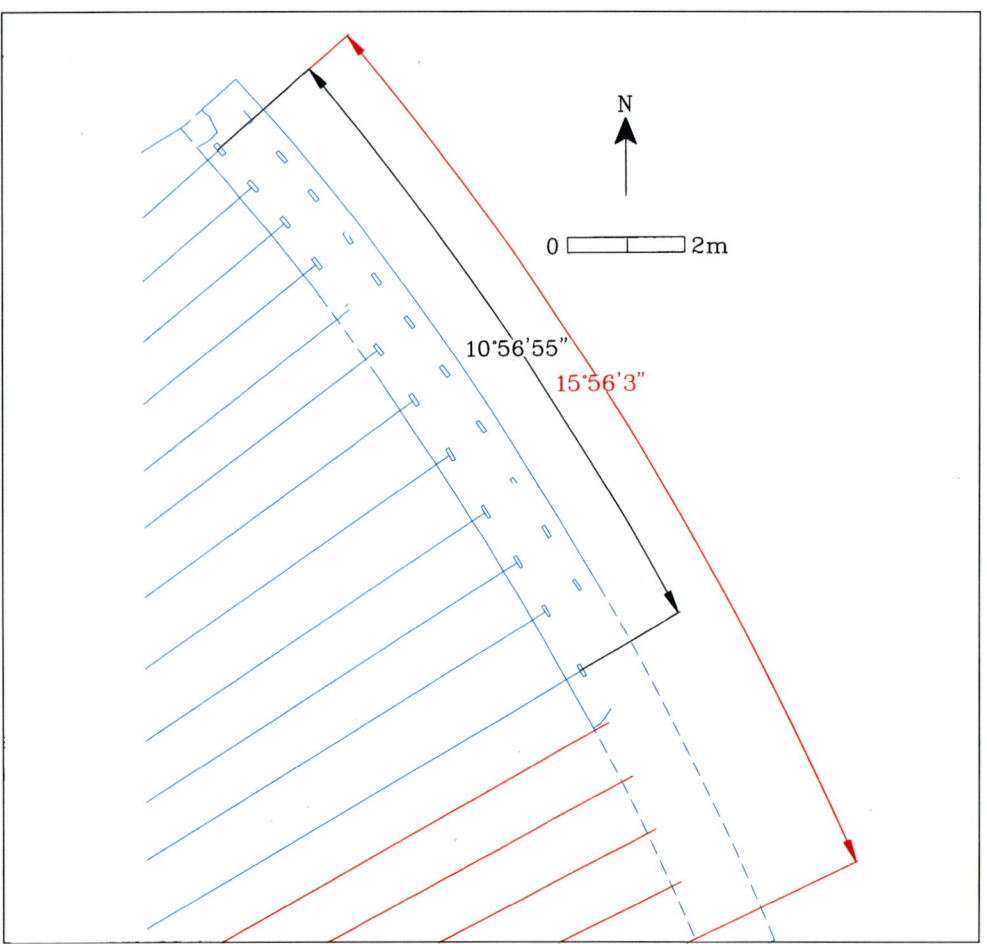

34. Detail of layout of Early Classical starting line.

eighth and the ninth starting position from the north end of the starting line is a small circular depression, 0.35 m. from the front of the starting line, 0.445 m. from position 8 and 0.455 m. from position 9. The depression is 0.015 m. wide and 0.005 m. deep and it may date from the original construction of the starting line since the blue-black pigment covers the depression. In addition, it may represent approximately the middle of the arc of the circle inscribed. From this central point on the starting line, the architect would have spaced out starting positions for the total number of starting athletes to be accommodated. The cord was swung to the north and to the south and at approximately every 0.951 m., center to center, which was almost exactly one degree of the circle (0d 59' 43") or 3.5 Corinthian feet, a starting position was set out (Fig. 33).[28] The actual lane widths at Corinth vary between 0.80 and 1.10 m. The first eleven lanes from the north end of the starting line at Corinth vary in angle between 0d 50' 23" to 1d 11' 28." The total arc between the first starting position and the twelfth is 10d 56' 55" and between the first position and the seventeenth is 15d 56' 3." It is likely that the total arc initially set out was intended to be 16 degrees for 17 positions, after which, for some reason, the individual positions were set out with less precision (Fig. 34). For comparison, at Isthmia, the lane widths of the fifth century B.C. *balbides* average ca. 1.05 m. It is most likely, based on the amount of available

space as well as the physical evidence, that eight positions would have been created to the north and eight positions to the south of the midpoint of the arc for a total of seventeen positions.[29] For each position, a front (left) toe groove and a rear (right) toe groove were provided as well as a painted number to designate lane.[30]

The resulting arc of the seventeen starting positions plus additional space to the north and south, adjacent to the first and seventeenth positions, equals 17d 15' 5" and the center of the arc lies at a point on the racecourse floor 55.274 m. (or ca. 200 Corinthian feet) from all seventeen of the starting positions and is approximately 6.30 m. from the south border and 10.26 m. from the north border of the racecourse. The width of the racecourse based on the width of the early starting line and the position of the Hero Shrine to the south and the quadriga base to the north has been restored as 16.56m.[31] The *dromos* is thus reconstructed as rectilinear in shape.

Each starting position was set out approximately one degree of this circle from each neighboring position. The total arc of the seventeen positions would equal approximately 16 degrees. The effect of this design, if employed in its simplest manner, would be to create a focal point at which all seventeen of the starting athletes would have run the same distance and converged, assuming that they had run at the same speed. However, the focal point of the starting line arc prob-

28. See Broneer, Isthmia II (1973): 49.

29. The section of the starting line that remains covered is between 3.80-4.70 meters.

30. The starting line as well as the letters representing lane divisions was resurfaced within the Classical period probably in order to raise its elevation

to the level of the racecourse floor and possibly to make its starting positions more regular. See Williams (1980): 10-11.

31. It should be noted that this distance of 16.56 meters as the width of the racecourse is very close to being one tenth of the length of the *dromos* (or 60 feet) if 165.82 m. is used as the calculated length of the course (55.274 X 3 = 165.82).

ably did not, in practice, serve as a point of convergence for the athletes but rather as a location of a "break post."

It is likely that a vertical post at the southern extreme of the "break line" would have been a visual reminder to the starting athletes to "keep to the right" and likely "stay in your lanes." The hypothetical location of the "break post" at Corinth is very good evidence that in Greek athletics the turns were made to the left and the resulting course to be run was counter-clockwise. In sculptural and painted representations, Greek runners are commonly depicted at the start of a race with their left foot forward. A good example of this is the miniature bronze athlete from Olympia (Fig. 2). This position would mean that the first foot to be moved from the starting line would likely have been the right. The necessity for this foot position is certainly the case in the two successive Corinthian starting lines based on the relative location of the toe grooves for the front left and rear right feet. As a consequence, the first movement would be toward the left and this may have influenced the counter-clockwise direction of the race and the turn, although there may have been a natural tendency of "right-footedness."

To the north of the "break post" would have been drawn a curved "break line," whose definition was an arc of a 100 foot circle with center near the absolute center of the track. This circle would have formed a tangent with the original circle having a radius of 200 feet (Fig. 35, 36). In practice this would mean that by the time that all of the athletes reached the curved break line in lanes, they would have run an equal distance. Clearly the least attractive lane assignments in this curved starting line system would have been the positions closest to the south end, *alpha, beta, gamma, delta*. This is due to the nature of the break post and "break line," which would have meant that the athletes in the southern lanes would have had to make a moderate change in the angle of their forward motion, to the left, after passing the break post. In order to carry out this maneuver they would have needed to slow their pace slightly to avoid running out of their lane and into another[32] Their lane widths would have diminished, from an average of 0.951 m. (3.5 Corinthian feet) at the start to 0.60 m. (2.2 Corinthian feet) at the break line after which the athletes would have been free to leave their assigned lanes and run toward the center of the track (Fig. 37). There may have been other rules. For instance, a prohibition against runners "cutting in" without sufficient space, as in modern track events. A necessary implication of this system is that the lanes be marked and visible on the racecourse floor. Otherwise it would be difficult for the athletes to stay within the limits of their assigned lanes. There is, however, no evidence for lane markings at Corinth, although at Nemea some evidence for coloring portions of the track surface have been found.[33]

In theory, therefore, by close to the absolute center of the racecourse, each of the starting runners would have been given a fairly equal chance to be at the same point in the race. In practice, many

32. The only way to have avoided this circumstance would have been to change the location of the turning post at the west end of the *dromos* from the approximate center of the width of the track to a location further to the north. This would have more greatly limited the width available to the runners on the north side of the *dromos* but could have compensated to a certain degree for the inequity of the southern lane assignments at the "break post." This appears not to have been done at Corinth, but may be responsible, at least in part, for the off-center turning post from the late fourth century stadium at Nemea. See below note 34.

33. See Romano, "Stadia" (1981): 86-88, fig. 54.

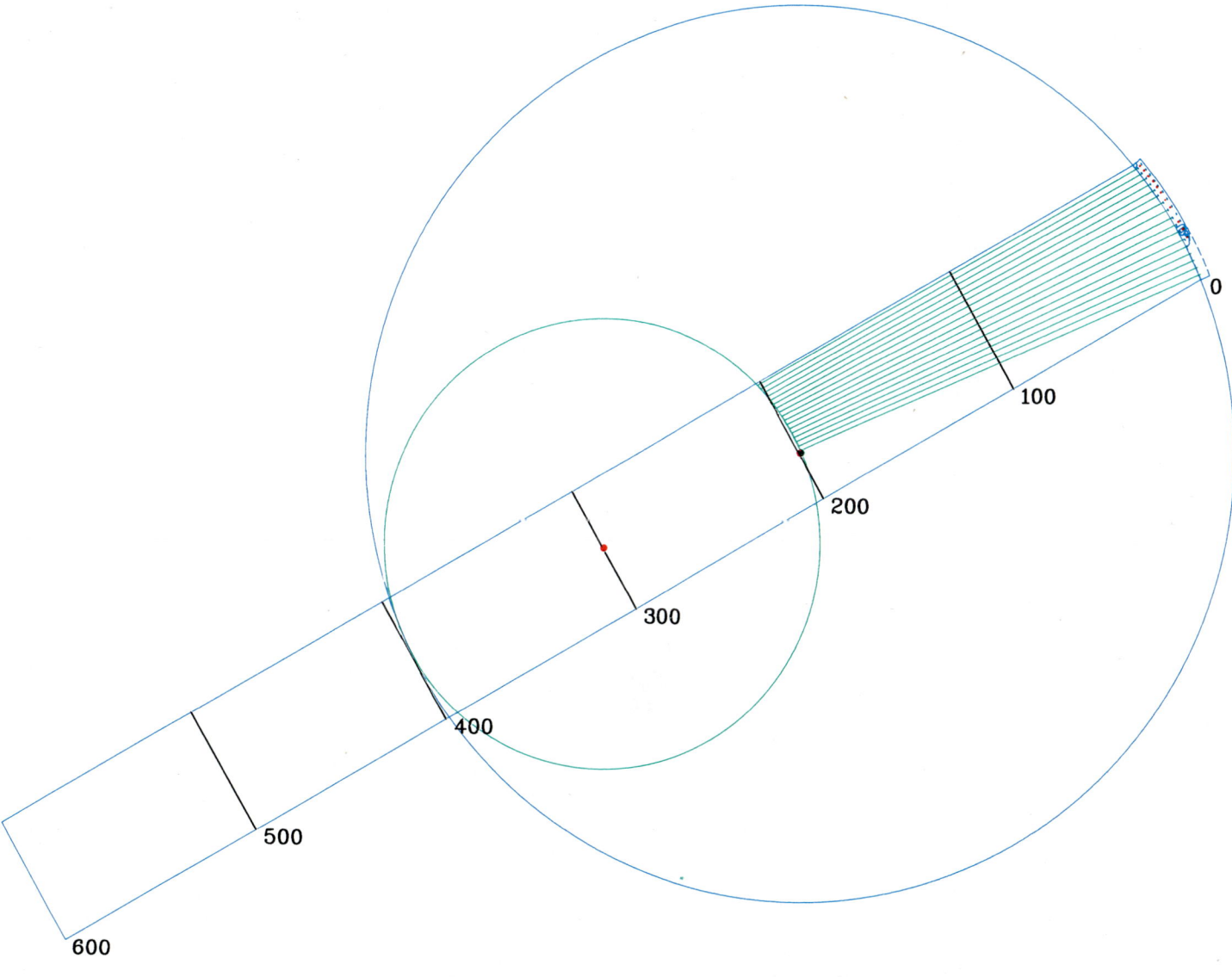

35. A tangent circle with radius of 100 feet is created with center near the absolute center of the racecourse. The arc of this circle creates the "break line" for the starting athletes at the 200 foot mark. The green lines indicate the actual paths run by the athletes from the starting line to the "break line."

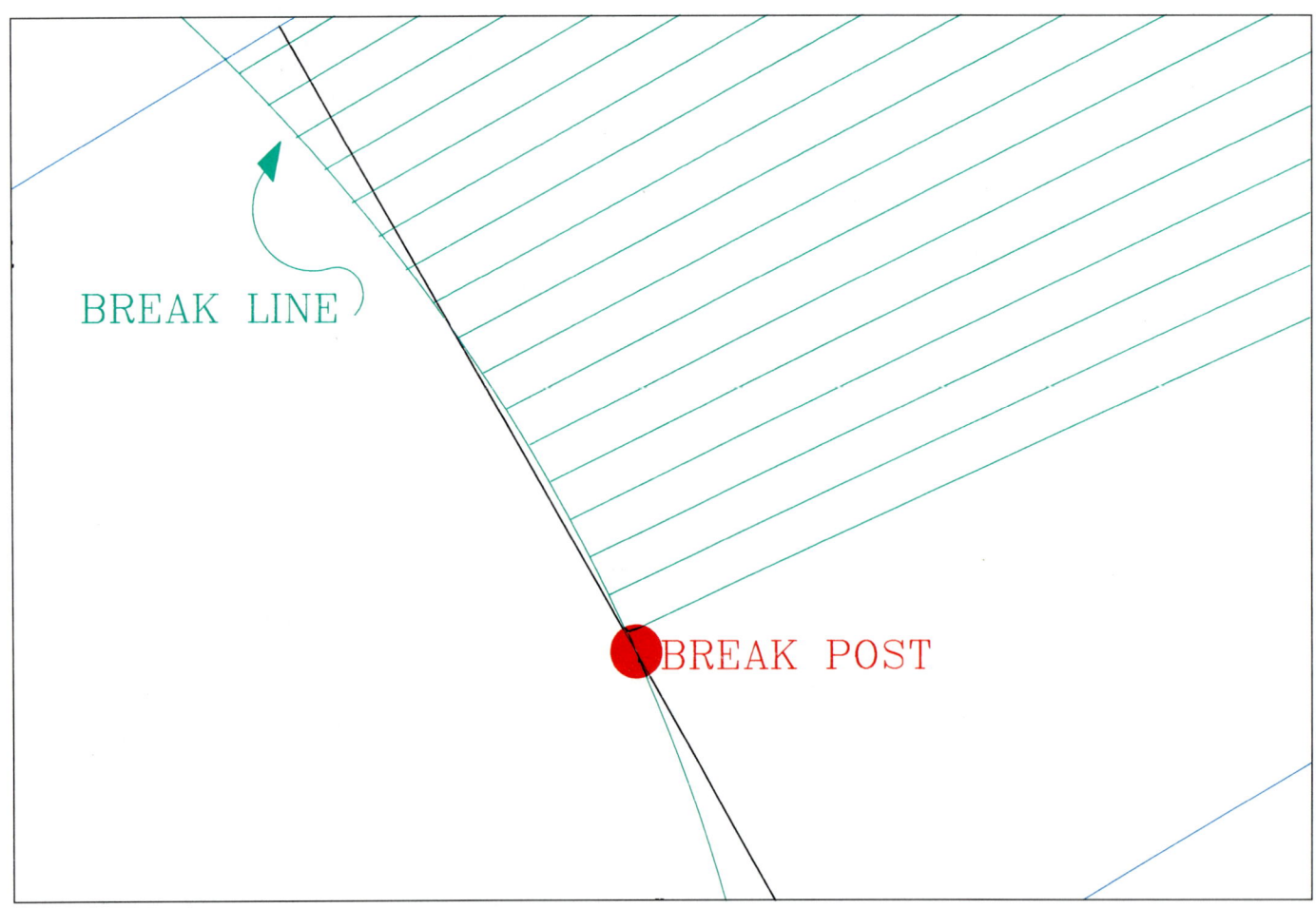

BREAK LINE

BREAK POST

36. Detail of focal point as "break post," and curved "break line."

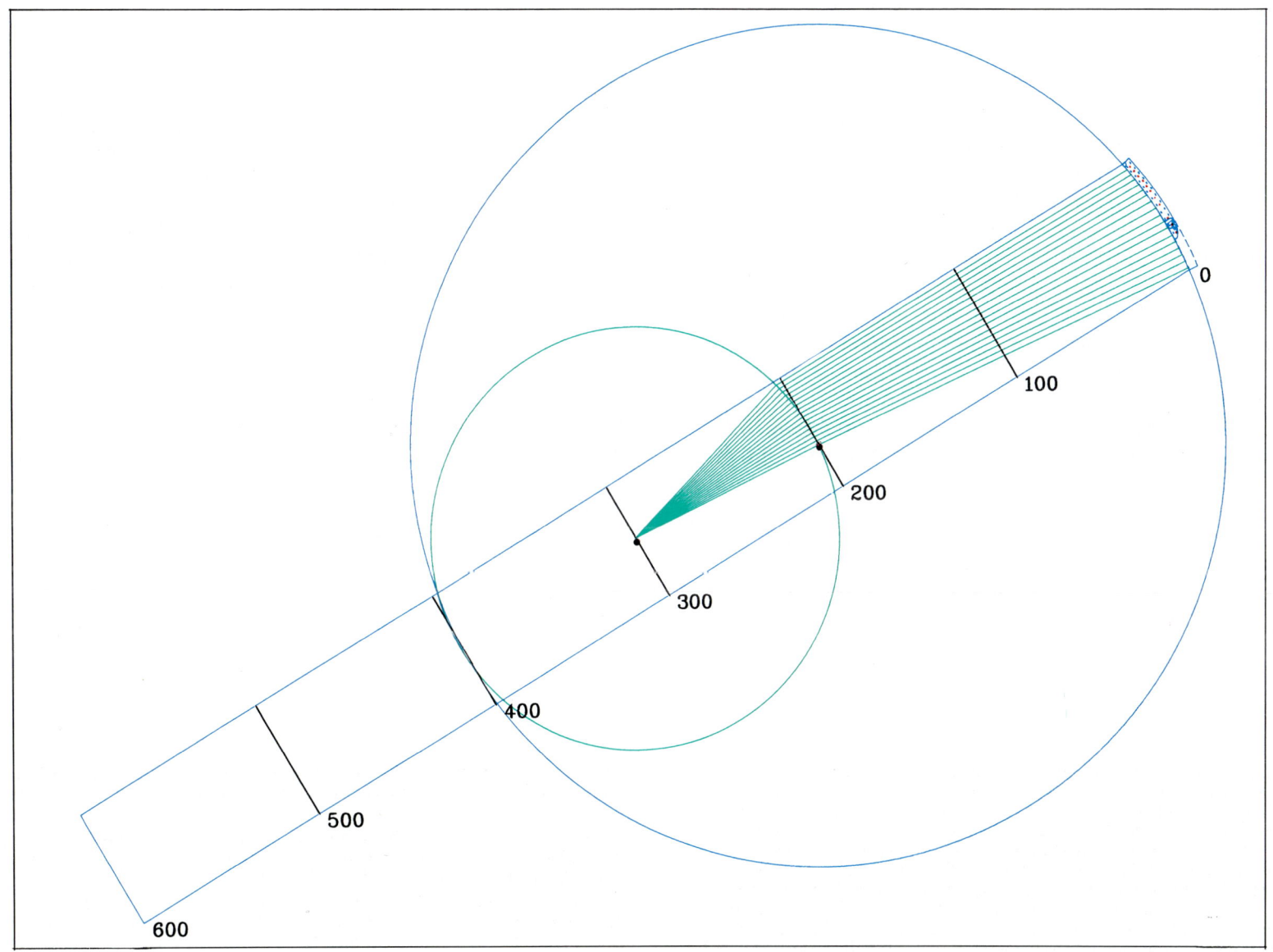

37. Equidistant lines to near the center of the racecourse, representing idealized paths run by athletes.

of the runners probably ran directly from the break line toward the distant west end of the racecourse where there would have been a single turning post (Fig. 38) around which the athletes would turn and return towards the eastern end of the course in a single file.[34] The race would continue in this fashion for the prescribed number of lengths of the course (Fig. 39).

The architect purposely positioned the focal point not in the center of the width of the racecourse floor but rather closer to the south side of the racecourse by 1.98 m. (7.2 Corinthian feet). The reason for this was to bring the starting runners who are spread out across the entire width of the racecourse at its east end, approximately 16.56 meters wide (60.2 Corinthian feet) to a location, about one-third of the available length of the racecourse to the west, where the athletes would be restricted to slightly more than one-half of the width of the racecourse floor, 10.26 m. (37.3 Corinthian feet) on the north side. The nature of the footrace itself would have necessitated that the athletes run not in parallel lanes but rather along the north side of the racecourse for its entire length,[35] turning at a turning post at the far

west end and then returning on the south of the racecourse for its full length. The obvious benefit of this would be the avoidance of head-on collisions that might have occurred if the runners were not forced to one side of the track at an early stage of the race. In itself, this design implies a footrace of at least four lengths of the racecourse, and possibly more (Fig. 39). A footrace of one length of the racecourse would have been run in parallel lanes. It is most likely that a footrace of two lengths of the racecourse would also have been run in parallel lanes, although probably two adjacent lanes would be assigned to each athlete.[36] The only known contests of greater than two lengths are the *hoplitodromos*, which was a race in armor of 2-4 lengths, the *hippios dromos* race of 4 lengths and the *dolichos* race which, from literary accounts, varied from 7-24 lengths of the *stadion*.[37] Unfortunately it is not known with which cults or dieties the starting line was associated.

There is some evidence to support the theory of the counterclockwise direction of the footrace at Corinth from an ancient contemporary source. There is a scene of a *dolichos* race on the reverse of a Panathenaic amphora by the Berlin Painter (New York, private

---

34. The stone base for such a turning post has been identified in the Nemea Stadium, located 5.30 meters in front of the southern starting line. The turning post is found off the central axis of the *dromos* towards the west by 3.40 meters. See Stephen G. Miller, "Excavations at Nemea, 1976," *Hesperia* 46 (1977): 22-25, pl. 16d. See also Romano, "Stadia" (1981): 104-108, where the suggestion is made that the "turning post" at Nemea may also have been a "finish post." In other stadia, it is often believed that one of the post holes in the stone starting line used for lane division could also have served as the receptacle for a turning post. In Corinth, a series of post holes have been excavated in the area immediately to the west of the ca. 500 B.C. starting line. It is possible that some of the post holes may have served as the receptacles for turning posts (in different years) for the returning runners. See Williams (1980): 17-18, note 24.

35. Since there are no post holes associated with the starting line, the turning post would have to have reduced the total length of the stadium as at Nemea (above note 34.)

36. See the discussion on this subject in R. Patrucco, *Lo sport nella Grecia antica,* Florence (1972): 106-110, and in Stephen G. Miller, "Turns and Lanes in the Ancient Stadium," *AJA* 84 (1980): 159-166.

37. For the distance of the *dolichos* see, Schol. Sophokles, *Elektra,* 687; Schol. Aristophanes *Birds* 291; AP IX, 319; Philostratus, *Gym.* 11; *Hippios,* Bacchylides IX, 25; Plutarch *Solon,* 23, 5; Euripides, *Elektra,* 826; Pausanias (VI, 16, 4); *IAG,* 44, 53, 56; Pausanias (VI, 13, 3.) The *hippios dromos* race was not commonly run at all athletic festival sites; it was known, however, at both Nemea and Isthmia.

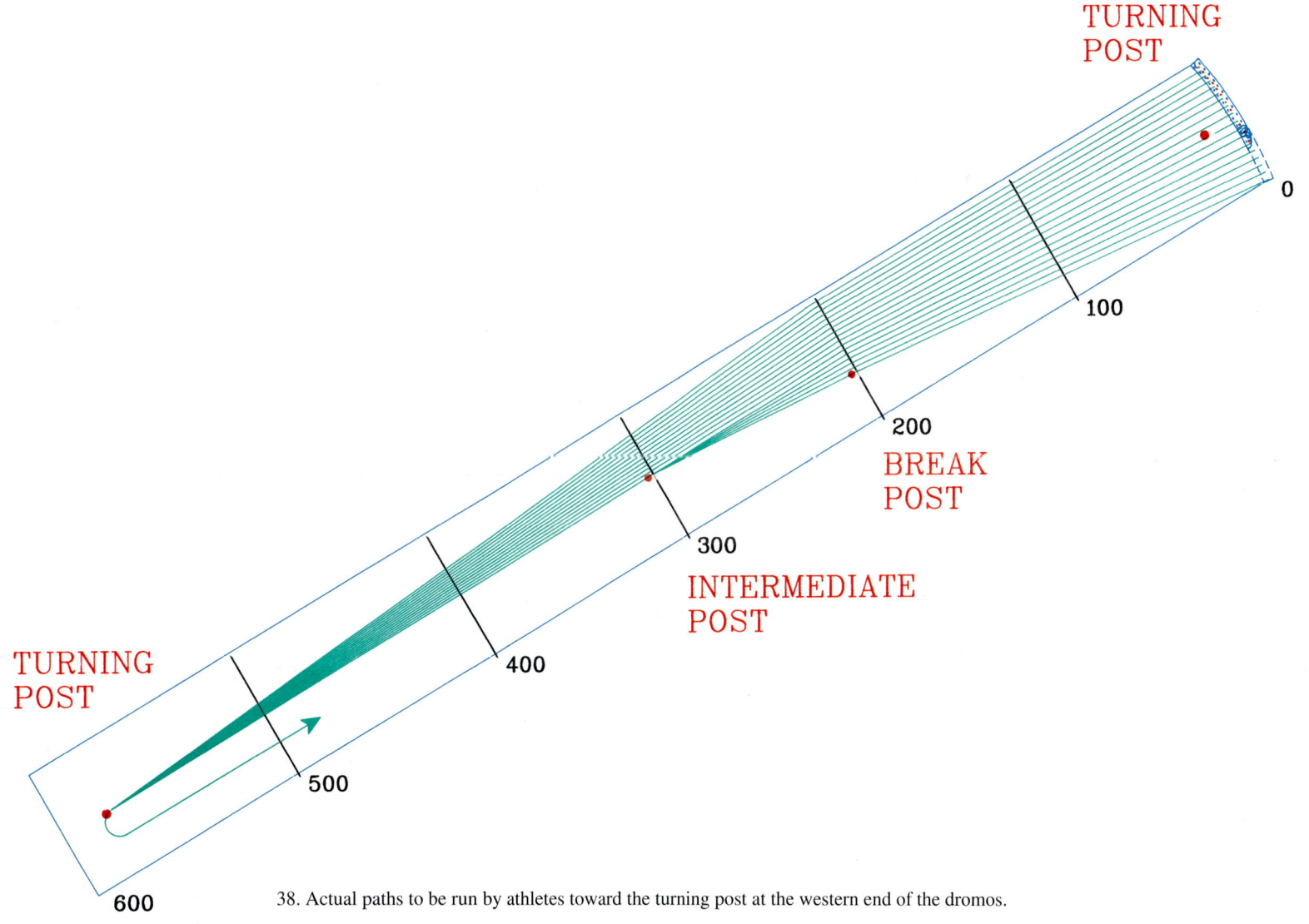

38. Actual paths to be run by athletes toward the turning post at the western end of the dromos.

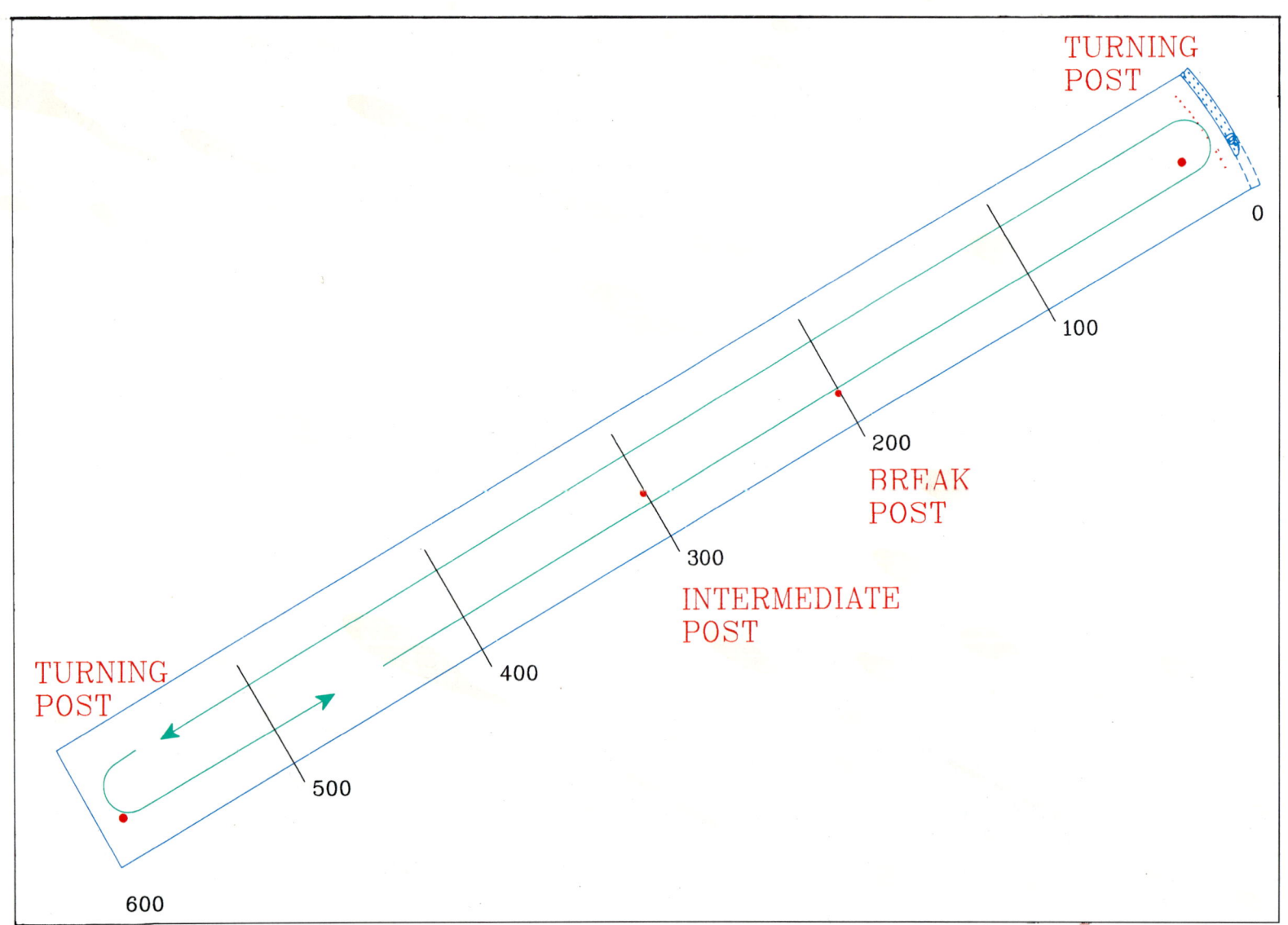

TURNING
POST

0

100

200

BREAK
POST

300

INTERMEDIATE
POST

400

TURNING
POST

500

600

39. Nature of the distance race, to be run around two turning posts.

collection; former Castle-Ashby) of ca. 480 B.C. that shows four tall runners approaching a vertical post, interpreted as a turning post or kampter and not a finish post (Figs. 40a,40b).[38] There is also the possibility that the visible post in the vase painting was an intermediate marker in the racecourse, as would have been required at Corinth. From the painted scene, one can clearly see that the lead runner's left and forward leg is passing on the right side of the post, thus suggesting that the turn would have been counter-clockwise. There is no indication from any excavated stadium that a spina or low barrier wall along the central axis of the racecourse, existed as it commonly did in the Roman circus. However, in the late fourth century B.C. stadium at Nemea, narrow bands of colored earth have been found set into the racecourse floor along the central axis.[39]

The Athletes

There has been scholarly discussion about the nature of the footraces run on the successive Corinthian racecourses. Morgan[40] suggests that the short grooves and the wide spacing of the foot positions indicated a starting position for athletes in a torch race, where the athletes would have taken a crouching start. He cites the Scholiast on Pindar's Olympia XIII, 56, who notes that a torch race formed a part of the festival in honor of Athena Hellotis at Corinth. Herbert[41] also believes that the early racecourse was used for the running of a torch race, citing four late fifth century B.C. Corinthian red figure bell krater fragments with representations of a torch race which were found in Classical levels nearby the racecourse. Herbert prefers to connect the contest with Artemis Hellotia instead of Athena Hellotis. The torch race theory seems less likely now since the discovery of the curved starting line. The starting athletes would be converging during the first third of the first length of the *dromos*, and the space between athletes with their torches in each lane would be reduced from a width of approximately 0.95 meters at the starting line to less than 0.60 meters by the "break line." The narrowness of the lanes by the "break line" would rule out the running of a torch race from this curved starting line based on reasons of safety and logistics. Williams[42] has suggested the possibility that the starting line could have been used for the *hoplitodromos*, the race in armor, although the narrowness of the lanes would have made this race difficult as well since the athletes would have been carrying large shields.[43]

Previous discussions of the successive starting lines in Corinth have assumed that the racecourses were designed for male athletic festivals and this is still most likely. It is also possible that females may have participated in cult footraces at Corinth, since female ath-

38. See J.D. Beazley, *The Development of Attic Black Figure*, Revised Edition, ed. by Dietrich von Bothmer and Mary B. Moore, Berkeley (1986): 187-188, note 45. I thank Professor John Keenan of the School of Engineering and Applied Science at the University of Pennsylvania for asking me the question "Why did the ancient Greeks run counter-clockwise, as we do in the modern day?"

39. See above note 33. The off-center location of the "turning post" at Nemea also suggests a counter-clockwise turn.

40. Charles H. Morgan, II "Excavations at Corinth, 1936-1937," *AJA* 41 (1937): 459.

41. Sharon Carey Herbert, "Corinthian Red Figure Pottery," dissertation in Classics, Stanford University (1972): 70-76.

42. Williams (1980): 13-15, note 20.

43. Some support for William's theory is found in the depiction of the start of a *hoplitodromos* race on a Panathenaic amphora of the mid-fourth century B.C. where three hoplite runners are depicted at the start of a race. The runner's feet are widely separated and are similar to the requirements of the Corinth starting lines. There is also depicted two horizontal ropes stretched across in front of the athletes, as a *husplex*, suggesting the starting arrangements at a straight starting line. See *Mind and Body* (1989): 251-252, No. 143.

40a. Panathenaic amphora, Berlin Painter, ca. 480 B.C. (New York, private collection, former Castle-Ashby). Obverse, striding Athena. Photos courtesy of the Metropolitan Museum of Art.

40b. Reverse, four runners approaching vertical post.

letic festivals are known from Greek and Roman antiquity.[44] There are two fairly well known miniature bronze representations of youthful female runners, British Museum bronze no. 208 from Yugoslavia, and Athens, National Archaeological Museum no. 24 (from the Karapanos Collection) from Dodona (Figs. 41, 42).[45] The British Museum piece dates to the late sixth century B.C. and the National Museum piece to the mid-sixth century. Both figures wear tunics, have long hair and are thought originally to be Lakonian. Their legs are widely spaced apart, left leg in front of right, and they have both been interpreted as being runners, running. They could also be runners at the starting line of a racecourse waiting for the start of the race. Each figure holds her short skirt with her left hand. Certainly this position would not be advantageous to a runner who would need to swing her arms vigorously during the contest. It would be more appropriate if it could be explained as a mannerism at the starting line. These miniature bronze figures of females, if at the start of a race, would be similar to the several known miniature male bronze statuettes of runners at the start of a race (Fig. 2). If these figures do depict female athletes at the start of a footrace then the Early Classical starting line from Corinth may be an example of a fairly common type of starting line from the sixth century B.C., possibly for both men and women.

Olympic victors from Corinth have been recorded in various sources.[46] There were at least 9 Corinthian Olympic victors between 776 B.C. and 146 B.C.:

Ol 13   DIOKLES, *stadion*, 728 B.C.

Ol 14   DASMON, *stadion*, 724 B.C.

Ol 67   PHEIDOLOS, horserace, 512 B.C.

Ol 68   _____, son of Pheidolas, horserace, 508 B.C.

Ol 69   THESSALOS, son of Ptoiodoros, unspecified footrace (*diaulos*?), 504 B.C.

Ol 77   [...]TANDRIDAS, *stadion* for boys 472 B.C.

Ol 79   XENOPHON, son of Thessalos, *stadion* and pentathlon, 464 B.C.

Ol 96   ———OS, wrestling, 396 B.C.

Ol 119   ANDROMENES, *stadion*, 304 B.C.

44. See, for example, Thomas F. Scanlon, "The Footrace of the Heraia at Olympia," *The Ancient World,* 9 (1984): 77-90 and also Thomas F. Scanlon, "Virgeneum Gymnasium," in *The Archaeology of The Olympics,* Raschke (1988): 185-216.

45. See Jane Sweeney, Tam Curry and Yannis Tzedakis, eds., *The Human Figure in Early Greek Art,* Athens (1987): 132-133. There is a third example in the National Archaeological Museum in Palermo, no. 8265. It dates to the third quarter of the sixth century B.C. and is illustrated in Doris Vanhove, ed., *Le Sport dans la Grèce Antique, Du Jeu à la Compétition,* Brussels (1992): 250, no. 114. This figure holds her skirt with her right hand. It should be noted that there are Bronze Statuettes of four male runners and a victor from the lid of a lebes depicting a similar position. They are British Museum GR 1824.4-89.2 a-e., and were found together with the Onomastos lebes from Campania, 480 B.C.; British Museum GR 1824.4 89.1.

46. They are included in the work of Luigi Moretti, *Olympionikai. I vincitori negli antichi agoni Olimpici,* Rome (1957) and in Luigi Moretti, "Supplemento al Catalago degli Olympionikai, *Klio,* 52 (1970): 295-339.

41. British Museum, bronze no. 208. Female runner, late sixth century B.C. Height 0.114 m. Photo courtesy of the British Museum.

42. Athens, National Archaeological Museum, bronze no. 24 (Karapanos Collection). Female runner, mid-sixth century B.C. Height 0.12 m. Photo courtesy of the National Archaeological Museum.

For chronological reasons, as well as due to the nature of the events, only four of these athletes could have used the Archaic *dromos*. In addition there are other known Corinthian athletes of the Classical period who could have used this racecourse. For example, from Simonides there is Nicoladas who was a famous athlete, and winner of the *stadion* and *pentathlon* but apparently, not an Olympic victor. From the Anth. Pal. 13.19,[47]

This statue is the dedication of Nicoladas of Corinth, who won the footrace (unspecified) at Delphi, who took sixty amphorae of oil in five prizes at the Panathenaia, whose rivals three times did not reach the oath-offerings at sacred Isthmus; who won three times at Nemea, four times at Pellene, twice at Lykaion and at Tegea, Aegina and rocky Epidauros, at Thebes and at Megara town; who at Phlius won the stadion and the pentathlon and made great Corinth to rejoice.

Of all Corinthian athletes, Xenophon, son of Thessalos, is probably the most famous. Pindar's Thirteenth Olympian Ode celebrates the double victory of Xenophon in the *stadion* as well as the *pentathlon* in the Olympic Games of 464 B.C.

47. *Lyr.Graec.* II,184.

Xenophon of Corinth,
Stadiodromos and *Pentathlon*, 464 B.C.

Praising a house
  that has won
    three times at Olympia,
a house
  gentle to citizens
    and thoughtful of strangers,
I will glorify
  Corinth, the blest,
    doorway to Poseidon's Isthmos,
brilliant in her young men,
  home of Eunomia
    and her sisters —
Dika, unshakable foundation of cities,
  and Eirena, preserver of wealth:
    golden daughters of sagacious Themis.
(1-8)

They are eager to repel
  Hybris, brash-tongued
    mother of Koros.
Yet there is beauty
  to tell of here,
    and boldness moves me
to tell it.
  Who can hide
    the nature with which he was born?

Sons of Alatas, on you
  the Horai rich in bloom
    have often showered
bright petals of victory
  in the sacred games, and often
    in the hearts of your men they cast
(9-16)

the seeds of ancient inventiveness: the glory
of every work goes to its maker.
Who first framed the dithyramb of Dionysos, sung
on the way to the altar where the ox is felled?
Who added the bit to the horse's gear
or set the eagle, king of birds, above
the temple pediment, at either end?
                    In this town,
the muses breathe sweetly, and Ares
bristles in the young men's deadly spears.
(17-23)

Father Zeus,
  ruling with wide sway
    from Olympia's height,
harbor no envy
  against my words
    now or ever.

Guide this people
away from harm,
and swell the sails
of Xenophon's luck.
Welcome the revelry
due him for his crowns.
He comes from Pisa's plain, winner —
as no mortal before him —
in stadiodromos and pentathlon too.
(24-31)

And twice
in the Isthmian games
plaited leaves of triumph
shaded his brow,
nor will Nemea
smart to hear him praised.
His father Thessalos
enjoys glory for speed
where Alpheos flows,
and honor at Pytho for taking
the single and the double race
within a single day,
and three times within a month
in rocky Athens, the day's
swift running gave him garlands,
(32-39)

and seven times Athena flung him wreaths
in her Corinthian games,
and in the rites
of Poseidon on the sea-fringed fields
his kinsman Ptoiodoros won, followed by
his sons, Terpsias and Eritimos: too long
a song will follow if I tell their deeds
at Delphi, or how many times they beat
opponents on the lion's pastures. Why try
to reckon all the pebbles of the beach?
(40-46)

In every matter
measure is the thing-
to know it
is all tact.
I have set sail,
one man
at a people's bidding.
Singing the mind of the ancients,
singing war,
I will not betray Corinth,
rich in heroic legend,
with Sisyphos, shrewd in devices
like a god, and with Medea,
who put her love before her father
to save the Argo and its crew.
(47-54)

And in the test of battle
  around Troy's walls
    on either side
Corinthian warriors appeared
  to determine the outcome —
    those who strove
with Atreus' dear sons
  to bring Helen home and those
    who beat them back.
Danaans trembled
  at Glaukos coming out of Lykia.
    Before them he boasted
of Corinth
  where his seat of power lay,
    the hall and splendor of his father
(55-62)

Bellerophon, who once had suffered much
beside Peirana's spring, deluded
in his need to catch and master Pegasos,
the son of snaky Gorgon —
                    until Athena
brought him the golden bridle, a dream at first
and then no dream: "Asleep, Aiolian king?
Here is a drug to calm a stallion's mood.
Rise and when you've killed a silver bull
in honor of Damaios, show him this."
(63-69)

The maiden,
  with the deep blue aegis
    before her in the dark,
appeared
  to speak these words.
    From the depths of slumber,
he jumped to his feet
  and found
    the marvel lying there.
In joy he went to Polyïdos,
  seer of the land,
    and showed him the bridle,
explaining all that had happened,
  how he lay night-long as the seer instructed,
    and how Athena came and gave him
(70-77)

the spirit-taming gold.
  "Obey the dream,"
    the seer replied,
"and when you've slain
  a bull for the Earth-Shaker,
    build an altar to Athena,
goddess of horses."
  The power of the gods
    brings to fulfillment —
as if it were a trivial thing —
  even the deed
    no man would promise

or expect.
  Strong Bellerophon hastened,
    slipped the soothing bridle over the cheeks
(78-85)

of winged Pegasos, and rode him
in maneuvers of war, mounted
in full bronze.
Then from the chill
folds of hollow heaven he let fly
the bolts of death, sweeping serried ranks
of Amazons, women armed with bows.
                              He killed
Chimaira, breathing fire, and the Solymoi.
His fall I won't recount, but Pegasos
entered Olympos and Zeus' ancient stables.
(86-92)

Now I must put strength
  in the throw,
    and let go
a whirl of javelins
  straight at the mark,
    for I have come
a willing ally
  to the muses
    and the sons of Oligaithos.
I will make their multitude of victories
  at Isthmos and Nemea
    manifest in few words:

sixty times they won at both.
  The crier's voice
    will vouch the truth of what I claim.
(93-100)

It seems I have
  already named
    their triumphs at Olympia.
Those that will come
  I would praise
    when they come.
I have my hopes,
  but the end
    is with the god.
If the family's luck endures,
  Zeus and Enyalios will do the rest.
    They won beneath Parnassos' brow,
at Argos and at Thebes,
  and in Arkadia the regal altar of Lykaian Zeus
    will testify to the host of their successes.
(101-8)

Pellana, Sikyon, and Megara,
the Aiakan grove of Aigina,
Eleusis, shiny Marathon, the towns
beneath high Aitna that abound in wealth,
Euboia too:
  search the whole of Greece
and you will find they've won too many times
to calculate.

*Come, then! swim out with agile strokes
and Zeus, Perfector, grant them reverence
and good fortune's sweet delight.
(109-15)*

From the translation by Frank J. Nisetich, *Pindar's Victory Songs,* Johns Hopkins University Press, Baltimore, 1980.

In this ode, not only is there considerable reference to the mythological traditions of Corinth (Sisyphus, Medea, Bellerophon and Pegasus) but there is also some detailed accounting of the athletic victories of Xenophon and his father Thessalos at the Olympic Games. According to Pindar, lines 29-31, Xenophon is the first man in the history of the Olympic Games to win both the *stadiodromos* and the *pentathlon* in the same Olympic Games. Winning both of these events at the Olympic Games certainly proves the athletic skill as well as the versatility of Xenophon. The *pentathlon* was the most eclectic of all the contests of Greek athletics, being composed of five different events; the footrace, most likely the *stadion*, long jump, discus, javelin and wrestling. Simonides[48] enumerates the events as ἄλμα, ποδωκείην, δίσκον, ἄκοντα and πάλην, "jump, swiftness of foot, discus, javelin and wrestling." Pindar also mentions (lines 32-34) that Xenophon has won twice in the Isthmian games and implies that he has won at Nemea as well. From another ode by Pindar (fragment below) we know that Xenophon pledged to the goddess Aphrodite on Acrocorinth that if he were victorious at Olympia, he would dedicate 100 courtesans to the service of her cult.[49] Xenophon made good on his promise and commissioned Pindar to write an ode (only a portion of which survives) to be sung in the sacred grove of the Temple of Aphrodite as the slave girls danced to the song.[50]

48. *Lyr. Graec.* II, 182.

49. This is reminiscent of the bronze dedication of Kleombrotos from Sybaris of the first half of the sixth century B.C. that reads, "Kleombrotos, son of Dexilaos, having won at Olympia, dedicated one tenth of his prize to Athena, of equal length and thickness, having vowed to do so." See *Revue des Etudes Grecques* 80 (1967): 569; *Supplementum Epigraphicum Graecum* 29.1017 and 35.1053.

50. There is also a similar account of this event in Athenaeus (XIII,573 F.)

For Xenophon of Corinth

Guest-loving girls! servants of Suasion in wealthy Corinth! ye that burn the golden tears of fresh frankincense, full often soaring upward in your souls unto Aphrodite, the heavenly mother of Loves. She hath granted you, ye girls, blamelessly to cull on lovely couches the blossom of delicate bloom; for, under force, all things are fair.

Yet I wonder what the lords of the Isthmus will say of my devising such a prelude for a sweet roundelay to be the companion of common women...

We have tested gold with a pure touchstone...

O Queen of Cyprus! a herded troop of a hundred girls hath been brought hither to thy sacred grove by Xenophon in his gladness for the fulfillment of his vows....

From the translation by Sir John Sandys, *The Odes of Pindar including The Principal Fragments*, Loeb Classical Library, Cambridge, Mass. (1968): 580-583.

Xenophon's father, Thessalos, won an unspecified footrace at Olympia in 504 B.C. (see above) and also won the *stadiodromos* and the *diaulos* at Delphi on the same day and won three unspecified victories at Athens within a single month (lines 32-39). Furthermore, Pindar cites seven victories (unspecified) of Thessalos at the games in honor of Athena at Corinth (lines 40-42). Pindar then goes on to mention numerous victories won by members of his family, Ptoiodoros, Terpsias and Eritimos at Delphi and Nemea (lines 42-46). Later Pindar enumerates sixty victories won at both Isthmia and Nemea by members of the Oligaithos clan (lines 93-100).

It seems clear that Xenophon is a member of a family known for its athletic accomplishments that must span, chronologically, the late sixth and the early fifth centuries B.C. It is probable that these athletes used the Early Classical starting line and the *dromos* in the Upper Lechaion Road Valley as a part of a cult festival in Corinth. The *dromos* may well have been used in connection with the curved terrace to the south of the starting line as the site of associated athletic contests. Whether this athletic festival is related to the games of Athena or not is impossible to know at the present time although it must remain a distinct possibility. Pindar praises Corinth for its athletic prowess and it is certainly possible that there were other famous athletes from Corinth of the same time period.

Chapter 3
GREEK MATHEMATICS

The nature of the reconstructed *dromos* in Corinth suggests an understanding of mathematics and geometry by the Greek architect that previously has been unrecognized as early as ca. 500 B.C. In order to explain the method of the design of the racecourse it is necessary to summarize briefly the general knowledge of Greek mathematical thought by this period.

By the end of the sixth century B.C. the science of Greek mathematics was well advanced. Ionian Miletus had produced many of the most important Greek philosophers of the sixth century who practiced geometry and mathematics: Thales, the founder of the Ionian School (640-546 B.C.), Anaximander (610-540 B.C.), Anaximenes (fl. ca. 546 B.C.) and Pythagoras of Samos, later of Kroton, of the second half of the sixth century.[1]

A good deal of our knowledge about the specific contributions of Thales to the science of Greek mathematics comes from the ancient author Proclus who was a Neoplatonist philosopher of the fifth century A.D.[2] Thales of Ionian Miletus is said to have lived in Egypt for a period of time and to have brought with him from Egypt a general knowledge of Egyptian geometry and astronomy.[3] It is very likely that Babylonian and Egyptian mathematics had a great influence on the development of Greek mathematics by the sixth century.[4] Babylonian mathematics had two number systems, a sexagesimal system, base 60, and a decimal system, base 10; the product of these two bases equals 600. In Akkadian, an eastern Semitic language of the Babylonians, the word kiššatu meant "totality." In Sumerian, šar

meant 3600 but also "totality." The number 3600 is the product of 60 x 60, 360 x 10 and 600 x 6.[5]

It has been said that Thales introduced the concept of abstract geometry, by which the relationship between different parts of geometrical figures could be studied and defined. The following general theorems are attributed to Thales: a circle is bisected by its diameter; the angles at the base of any isosceles triangle are equal; if two straight lines cut one another, the vertical and opposite angles are equal; an angle inscribed in a semicircle is a right angle.

The geometrical schemes employed at Corinth in the laying out of the starting line were not very complicated but they were based on an understanding of the basic principles of plane geometry. The only real necessities were accurate measures of 100, 200 and 300 feet, the ability to swing the arc of a circle with a cord of those measurements from the same center, and the knowledge of the meaning of and relationships between the radius, diameter and circumference of a circle of 360 degrees.

Before the discovery of the curved Corinthian starting line and this study, it was not attested that sixth century B.C. Greek mathematicians were aware of π, the mathematical relationship between the diameter of a circle and its circumference.[6] But it will become clear in the following discussion that the Greeks must have had a value for π and employed it at Corinth. Also previously unattested is the fact that the sixth century Greeks knew that the circle was composed of 360 degrees.

---

1. His proof of the "Pythagorean Theorem" was likely his own contribution to Greek mathematics, however the fact of the theorem was known to the Babylonians from very early times. See Neugebauer (1969): 29-52.

2. See Heath, 1921, especially chapter IV, "The Earliest Greek Geometry. Thales," pp. 118-140 and Neugebauer (1969): 71-96.

3. I thank Professors Åke Sjöberg and Erle Leichty of the Babylonian Section of The University Museum for discussing with me Sumerian and Babylonian questions.

4. Herodotus (2,109) suggests that the Greeks learned the practice and methods of measuring land from the Egyptians but the use of the sun clock, the

sundial and the measurement of the day into twelve parts from the Babylonians.

5. Is it possible that the Greek *dolichos* race, which was the longest footrace in Greek athletics, was originally 7 lengths of the *stadion* and, taking in consideration the turning posts and the amount of the 600 foot racecourse that would have been diminished by the turning athletes at each end, the actual length run would be 6 x 600 = 3600 feet?

6. It is acknowledged by historians of mathematics that this relationship was expressed and understood by Archimedes in the third century B.C. See below, pp. 78-79.

Egyptian and Babylonian mathematicians knew of π and each had a numerical equivalent for it. A group of mathematical clay tablets from the Old Babylonian Period, excavated at Susa in 1936, and published by E.M. Bruins in 1950, provide the information that the Babylonian approximation of π was 3 1/8 or 3.125.[7] From the Rhind Mathematical Papyrus discovered in Thebes from the Hyksos Dynasty of 1650 B.C. and now in the British Museum, comes an approximation of π as $(16/9)2 = 3.1605$.[8]

There are at least two ways in which the Greek architect could have laid out the starting line at Corinth with starting positions one degree apart, on average. The first way would be to conduct a physical survey of the area from the focal point with an instrument of sufficient accuracy to discriminate one degree increments. Since we know virtually nothing about Greek survey instruments, it is difficult to know definitively whether the Greeks could have achieved this. It seems unlikely, however, since the *Corpus Agrimensorum Romanorum,* a collection of fragments of Roman surveyors' manuals including techniques and procedures does not include the description of an instrument that would have been capable of such a procedure.[9] In the Roman period, the known surveying instruments were employed to create straight lines and right angles only. Hero of Alexandria, in the first or second century A.D., describes a kind of sophisticated survey device, the *dioptra,* which he claims not to have invented, but to have perfected. The *dioptra* does not appear to have been an easily transportable instrument.[10]

The second way of laying out the starting line at Corinth would have been to determine mathematically the circumference of the circle with a known radius by employing the formula $C=\pi d$. In order to do this the Greek architect would have had to have used a numerical value for π. We can only work backwards toward the reconstruction of this ancient value for π, with only an approximation for the diameter based on the statistical computer model for the radius of the circle between the starting line and the focal point.[11] Using the formula $C=\pi d$, the following is known about the Corinth starting line where C=circumference and d=diameter. Since the average distance between starting positions is 59'49" or approximately one degree of the circle, which equals a chord length of 0.951 m. (3.5 Corinthian feet), if we multiply 360 x 0.951 m. (3.5) we have as a result 342.36 m. or 1260 Corinthian feet as the circumference of the circle. We have d= (2 x radius length of 55.274 m.) or 110.548 m. Dividing d/C = 3.0969 which is the value of π for this example. This value can be rounded to 3.10. The difference of chord length from arc length for a one degree measurement in a circle with radius 55 m. is negligible. The only factor that could change this estimated value for π would be to determine from archaeological excavation the actual location of the focal point of the starting line, if it should still exist, which could change the value of the diameter (d). For instance, substituting in the same formula, a shorter diameter (radius = 54.488), the value of π would equal 3.1416.[12] The percentage of error between these two values is 1.4 percent. It is known that Archimedes in the third century B.C. had a

---

7. Neugebauer (1969): 46-47.

8. Arnold Buffum Chace and Henry Parker Manning, *The Rhind Mathematical Papyrus,* Volume II, Oberlin (1927): 35-38. For an interesting account of 4000 years of π, see Petr. Beckmann, *A History of Pi,* Boulder, Colorado, 1982.

9. *Corpus Agrimensorum Romanorum,* i, ed. C. Thulin, Teubner, Leipzig, 1913. For a brief summary of the contents of the *corpus* see O.A.W. Dilke, *The Roman Land Surveyors, An Introduction to the Agrimensores,* Newton

Abbot (1971), Appendix A, pp. 227-230.

10. Hero of Alexandria, Volume III, *Metrica, Dioptra,* Teubner Edition (ed. H. Schone, 1903). See also O.A.W. Dilke, above note 9.

11. See above pp. 46-62.

12. This difference in radius length could also be accounted for if instead of measuring from the front incised line of the starting line, the measurement was made from a location slightly in front of the starting line, perhaps the first point where the right foot of the starting athlete touched the *dromos.*

value of π in the interval between 3.14286 and 3.14084.[13]

By the time of Ptolemy in the second century A.D., it is known that he had a value for the ratio of the circumference of a circle to its diameter expressed in sexagesimal fractions equal to 3.1416. Ptolemy is also known to have had a table of chords giving the length of chords of a circle, in relation to various diameters, subtended by arcs of 1/2 degree, 1 degree, 1 1/2 degrees etc. Hipparchus of the second century B.C. is said to have calculated a table of chords, although the actual work is lost. It would seem a good possibility that the sixth century B.C. Greeks also had such a table of chords available to them and that the choice of one degree increments for the starting positions at Corinth was no coincidence. The creation of a table of chords would imply a practical or theoretical knowledge of trigonometry which, in the Greek period, is usually first ascribed to Hipparchus in the second century B.C.

It is difficult to point to other attested examples of similar uses of the geometrical principles of radius, diameter and circumference in Greek architecture or engineering before ca. 500 B.C.[14] The curved starting line at Corinth would appear to be the earliest known example from the Greek world that presupposes an understanding and demonstrated use of these geometrical relationships and principles. It is well known, of course, that monumental Greek Doric architecture began in the Corinthia in the seventh century B.C. as did the development of the terracotta roof tile and the antefix.[15] Corinthians were known as inventors in other disciplines as well, including the introduction of black figure vase painting. It would not be surprising to learn that Corinthians were leaders in the practical application of geometry.[16]

13. Heath (1921): 232-235. See also Wilbur Richard Knorr, *The Ancient Tradition of Geometric Problems,* Boston (1986), chapters 1, 5.

14. There are, of course, the well known architectural refinements of certain Archaic, Classical and Hellenistic Greek temples which can be, and have been, analysed mathematically. See, for instance, the work of William Henry Goodyear, *Greek Refinements: Studies in Temperamental Architecture,* New Haven, 1912. An early example of these refinements may be found in the Archaic Temple of Apollo at Corinth, ca. 540 B.C. See William Bell Dinsmoor, *The Architecture of Ancient Greece,* reprint of 1950, Third Edition revised, New York (1975): 165-176. For a recent mathematical discussion of certain architectural refinements at Didyma see Lothar Haselberger and Hans Seybold, "Seilkurve oder Ellipse? Zur Herstellung antiker Kurvaturen nach dem Zeugnis der Didymeischen Kurvenkonstruktion," *Archäologischer Anzeiger* (1991): 165-188.

15. Major archaic limestone temples were built both in Corinth (see cover photograph) and in Isthmia. Examples of hinged bronze architectural instruments, calipers or compasses, have been found in the Archaic fill of the Temple of Poseidon at Isthmia suggesting their possible use in the design and construction of the temple. See Oscar Broneer, "Excavations at Isthmia, Fourth Campaign, 1957-1958," *Hesperia* 28 (1959): 329-330. These bronze instruments from Isthmia will be included in a forthcoming general study of the Isthmia bronzes by Isabelle Raubitschek. I thank A.E. Raubitschek for this reference.

16. See, for instance, Edouard Will, *Korinthiaka,* Paris (1955): 572-582; J.B. Salmon, *Wealthy Corinth,* Oxford (1984): 101-127.

Chapter 4
THE STARTING LINES FROM THE CLASSICAL *STADION* AT ISTHMIA

Only eleven kilometers to the east of the Corinth *dromos* is the Panhellenic Sanctuary of Poseidon that was under the supervision of the city of Corinth for most of its history. As presented above in Chapter 1, there is a contemporary stadium at Isthmia and the design and arrangements of the starting line should now be considered for comparison to the Early Classical starting line at Corinth. What possible relationships are there between these nearly contemporary facilities?

The starting gates of the Earlier Stadium, dating generally to the fifth century B.C., have been restored by Broneer, tested repeatedly and have been found to function well mechanically (Figs. 43a, 43b).[1] For each of the sixteen gates, a horizontal wooden bar is loosely hinged to a vertical wooden post, and has attached to it a cord which runs down the length of the vertical post and passes through a groove in the stone pavement. The cord is held in place in the groove by bronze staples and leads to a centrally located starter's pit where the starter would hold the other end of the cord. Theoretically, the starter could start all sixteen runners at the same time by simultaneously releasing the sixteen individual cords. Although it appears to the naked eye that all the gates fall at the same instant, in fact the starting gates fall at slightly different times (as can be seen in photographs) since the distance between the individual starting gates and the starter's pit ranges from approximately one meter in the inside lanes to approximately ten meters in the outside lanes.

It has been estimated that the velocity of each of the sixteen cords is approximately 100 meters per second and that consequently the starting gates in the outside starting positions will fall 1/10 of a second later than the gates in the inside starting positions, if the cords are released at the same time.[2] The formula employed to determine this estimate is

$$\text{velocity} = \sqrt{\frac{\text{tension}}{\text{mass per unit length}}}$$

where tension equals 1 kilogram x 9.8 and where mass per unit length equals 10^{-3}. Such a difference of 1/10 of a second would be significant in modern track races, especially in races of 400 meters and less which are measured to hundredths of a second and in which winners are often decided by the same margin.[3] Since it is most likely that the starting gates at Isthmia were used for the *stadion*, the shortest footrace of one length or the *diaulos*, a footrace of two lengths of the *dromos*, a distance between 300 and 400 meters, the starting gate system could definitely have been a factor in the outcome of the competitions. If released at the same moment, the sixteen starting gates would have always given an advantage to the athletes starting from the inside lanes, with greater disadvantage to each successive lane toward the outside. It is not known for certain if the inequity of the *balbides* starting gate apparatus at Isthmia was recognized in antiquity and whether it was responsible for the very short life of the starting gates. It is known that after only a brief period, possibly ten to fifteen years, according to Broneer, the triangular pavement was replaced by a different starting system. Broneer suggests that the *balbides* proved

---

1. For a discussion of the working of the starting gates see Broneer, *Isthmia* II (1973), Appendix II: 137-142, ΒΑΛΒΙΣ, ῩΣΠΛΗΞ, ΚΑΜΠΤΗΡ.

2. This must remain only an approximation since there are a number of variables that are difficult to reproduce accurately in the modern day, including the exact size and weight of the horizontal bar and the weight and texture of the cord.

3. I thank Professor Thomas H. Wood of the Department of Physics of the University of Pennsylvania for taking the time in 1980 and again in 1991 to discuss this problem with me.

43a. Isthmia, Photograph of the *balbides* sill with restored wooden gates. Photos courtesy of the University of Chicago, Isthmia Excavations.

43b. Isthmia, Photograph of starting athletes at the restored *balbides* sill.

unnecessarily cumbersome and were replaced by a simpler starting line. It is also possible that the starting gates were found to be unsuccessful for the reason cited above and that a different and simpler system was installed.

In the Early Classical starting line at Corinth the individual toe holds for the feet of the runner guaranteed a specific starting position at the starting line and one which would not have compromised the nature of the start. With an athlete's legs spread apart, as they would have had to have been at Corinth, it would be difficult to start undetected before the signal. In addition, because of the wide stance, it is likely that it would have been easier for an athlete to balance at the starting line. Although it might be argued that the distance between toe grooves at Corinth is unnecessarily great, it may have been for the reason of stability that the space was so wide. In addition, the individual toe grooves guaranteed a specific starting location so that there was no need for vertical posts as lane dividers, as became commonplace in later Greek and Roman stadia.

At Isthmia, the *balbides,* by their design, did not require any specific foot position, although there may have been regulations about this that have not survived. The only requirement implied by the existing balbides is that the athlete stand behind the horizontal bar, but since there were no toe grooves it would have been possible for an athlete to lean too far forward and lose his balance before the race began or to move his feet behind the gate before the start.

It seems that the Corinthians at Isthmia and Corinth were experimenting with designs for starting lines for both short and long distance footraces. With the combined evidence of the two different excavated examples we have an idea of what a complete Late Archaic or Early Classical racecourse might have looked like: for the distance race, a curved starting line at one end of the racecourse and for the short races a straight starting line at the opposite end.

It can be assumed that the *balbides* at Isthmia were installed in the fifth century largely as an attraction for spectators watching the athletic contests there. Broneer points out that the word *balbides* was well known in the fifth century, since Aristophanes uses the word twice in his extant plays.[4] It is possible that similar starting gate systems may have been used at other athletic sanctuaries, but Athenian audiences may have been familiar with the *balbides* gates from Isthmia. Only after some degree of use was it realized that the starting gates of the type at Isthmia did not, in fact, provide for a fair start. The life of the Isthmian *balbides* was relatively brief whereas the life of the starting lines in Corinth was very long, over 300 years. This would support the notion that the starting lines in the Upper Lechaion Road Valley, both curved and straight, were found to be satisfactory and thus were retained over time. It was a simple design and it guaranteed a reasonably fair start for the athletes. The fact that other starting line blocks with individual toe grooves (although more closely spaced) have been found at Nemea would support the conclusion that this was a successful design used for other stadia.

At Isthmia, the *balbides* starting gates were replaced by a single grooved starting line (above Chapter 1) by the beginning of the fourth century B.C. The single groove provided a location for at least one foot of the athlete.

---

4. Broneer, *Isthmia* II (1973): 139-140, *Knights* 1159; *Wasps* 548. The *Knights* was first produced in 424 B.C. and the *Wasps* in 422 B.C. It is possible, based on Broneer's evidence, that the *balbides* sill was introduced in the Earlier Stadium at Isthmia in the second half of the fifth century and even possibly the last quarter.

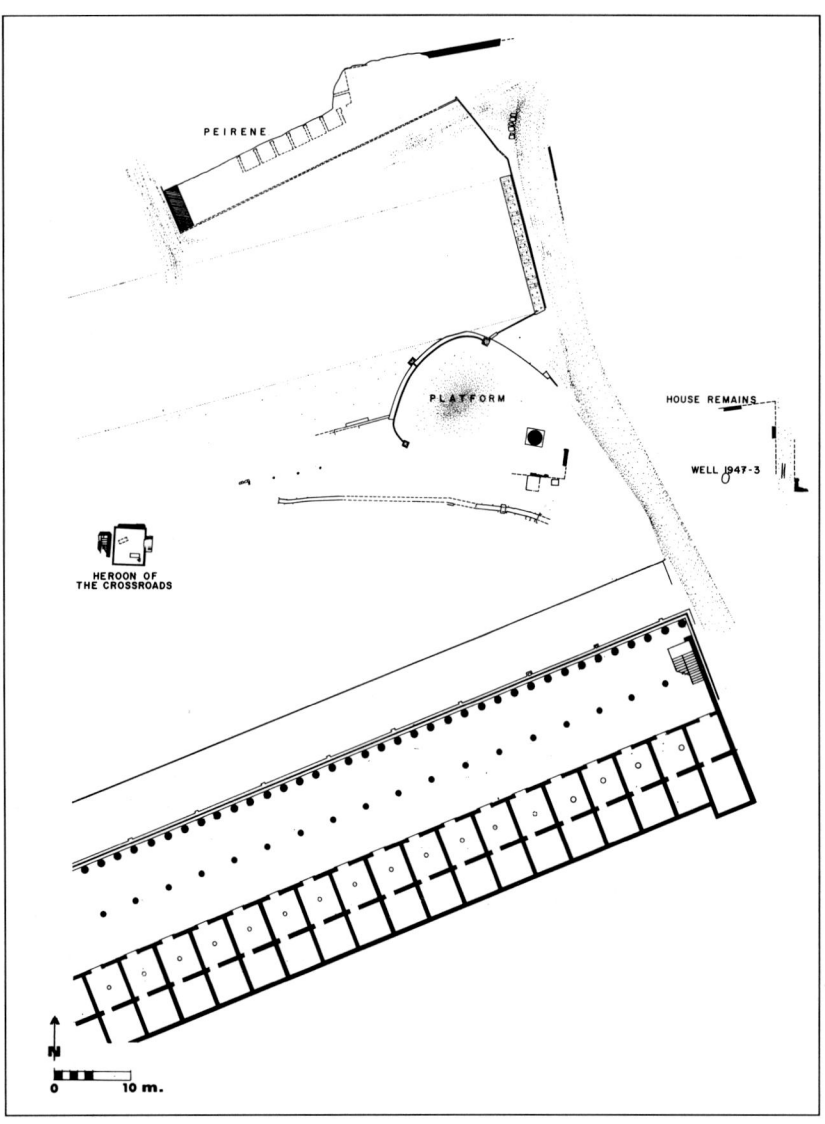

44. Corinth, East end of Hellenistic *dromos* with South Stoa, ca. 250-225 B.C.
Courtesy of the American School of Classical Studies at Athens, Corinth Excavations.

Chapter 5
THE HELLENISTIC *DROMOS* IN CORINTH

In the years after the construction of the South Stoa at Corinth, in the last third of the fourth century B.C., limiting the south side of the Upper Lechaion Road Valley (Fig. 44) the Archaic *dromos* was replaced by a Hellenistic *dromos* which was reoriented approximately eighteen degrees, to a more east-west direction (Fig. 31).[1] The Hellenistic starting line was constructed slightly to the east of the Early Classical starting line, overlapping and destroying the southern five starting positions of the earlier line (Fig. 45).

The straight starting line of the Hellenistic racecourse measures 17.2 x 1.35 meters; with the added stone projections at the north and south ends of the starting line the total length is 20.41 m.[2] The starting line is built of long poros blocks, 1.04 x 0.44 m. set in rubble and plaster and covered with fine cement.[3]

In the first phase of the starting line, starting positions for seventeen athletes were cut into the poros blocks. Each position was composed of individual toe holds, approximately 0.18 -0.20 meters long, 0.07 m. wide and 0.56 m. apart, measured center to center. Similar to the toe grooves of the early classical starting line, each groove has a vertical back wall and a beveled front wall. The maximum depth of the grooves is 0.05 meters. The front groove is positioned slightly to the left of the rear groove. The individual starting positions are spaced at intervals of 1.03 m.[4]

In the second phase of the starting line, the central starting position was covered over with cement to provide a blank position, flanked by two groups of eight positions.[5] Immediately adjacent to the center of the blank position, on its west side, is a poros block with a cutting 0.22 x 0.21 m. in its top surface. A second cutting joins this poros block with a second post hole cut into the racecourse floor and found immediately to the west of the block. A large post hole is located 11.21 m. to the west of this block and may be related to the running of the races (Figs. 24, 46, 47).[6] Probably contemporary with the above changes of the starting line was the addition of a large stone block at each end of the line. These blocks, 1.407 x 0.630 m. at the north end and 0.958 x 1.098 m. at the south end, have cuttings in their top surfaces, possibly related to the centrally located poros block. These blocks were likely used as foundations for the mechanical device, *husplex*, that regulated the start of the footraces. Similar stone blocks have been found as a part of the starting lines of the stadia at Epidauros, Nemea and the later stadium at Isthmia. *Husplex* blocks reused in the *Hellanodikai* area of the Olympia stadium have also been identified.[7] Broneer first identified the blocks in the Epidauros stadium as a part of a *husplex*, an ancient Greek word meaning "something triggered to spring open or shut."[8] The source of the information about the *husplex* at Epidauros is a third century B.C. inscription

---

1. For a complete discussion of the excavation of parts of the Hellenistic racecourse see, Charles K. Williams, II "Corinth, 1969: Forum Area," *Hesperia* 39 (1970): 1-39 and Charles K. Williams, II and Joan Fisher, "Corinth, 1970: Forum Area," *Hesperia* 40 (1971): 1-51.

2. This measurement comes extremely close to the width of the *balbides* triangular sill at Isthmia which measures 20.42 m.

3. No painted letters have been found on the surface of the Hellenistic starting line although it is likely that they would have originally existed.

4. This measurement is close to the average distance between vertical posts (lane width) at the *balbides* sill at Isthmia of 1.05 m.

5. This arrangement is also similar to the *balbides* line at Isthmia.

6. Charles H. Morgan, II, "Excavations at Corinth 1936-1937," *AJA* 41 (1937):

549-550, plate 16 reports that connecting these two holes was found a channel of irregular width and depth cut through the floor. It is possible that this large hole held a vertical wooden post as a turning post, although it may have been originally a part of the first phase of the Hellenistic starting line. This is suggested because the large hole to the west of the starting line is found slightly off center of the long axis of the racecourse, 8.186 m. from the south limit of the track and 9.157 m. from the north limit. It is also possible that the vertical post could have been a finish post for the races.

7. See above Chapter 1, p. 24 , note 50 and Romano, "Stadia" (1981): 140 note 21.

8. Broneer, *Isthmia* II (1973): Appendix II: 137-142, ΒΑΛΒΙΣ, ΎΣΠΛΗΞ, ΚΑΜΠΤΗΡ. pp. 137-142.

from the Sanctuary of Asklepius, IG IV² 1.98, which mentions a fine of 500 drachmai charged against Philon, a Corinthian engineer of the *husplex*. Apparently Philon had not met the stipulated conditions of the contract and subsequently was fined. It is possible that the same man, Philon, who worked on the *husplex* at Epidauros had earlier designed and installed the *husplex* at Corinth. At least the date in the third century B.C. agrees with the date after which the Hellenistic *dromos* was constructed (after 270 B.C.). The *husplex* appears to have been a later addition at Corinth, but this is still possible within the third century B.C.[9]

The Hellenistic racecourse was, in large measure, artificially constructed of earth fill which was brought in and deposited, especially in the area immediately south of the Temenos of the Sacred Spring (Fig. 31). The filling operation is dated to after 270 B.C. on the basis of numismatic evidence.[10] The racecourse floor, composed of a series of crushed poros layers, slopes downwards between 0.50-0.75 meters from the eastern starting line to a minimum height near the mid-point of the racecourse where the track surface ascends to the west.

The north side of the Hellenistic racecourse was clearly defined by the south side of the Peirene fountain house, by the triglyph wall of the Temenos of the Sacred Spring and by the monument bases which were erected on that wall, as well as by additional monuments that were set up to the east and west of the wall.[11] The south side of the racecourse was bordered by a quadriga (four-horse chariot) base, which was constructed earlier than the racecourse and was a feature to the north of the Archaic *dromos*, and by a second monument, probably for a *biga* (two-horse chariot).[12] The western end of the Hellenistic racecourse is as yet unexcavated, but its length is likely to be the same as that of the Archaic *dromos*, ca. 165 meters. The western limit of the Hellenistic racecourse would, therefore, lie beneath the Babbius Monument and the Fountain of Poseidon, two structures at the west end of the Roman forum.[13] It has been suggested by Williams that there may be cuttings for the foundations of a spectator embankment in the general area of what later became the Roman Central Shops, to the south of the *dromos*, but this is by no means certain.[14]

A stone water channel and six water basins have been discovered along a stretch of the southern side of the Hellenistic racecourse.[15] A Hellenistic reservoir, in the plan of two interlocking rectangles, is on an almost direct line with the six extant water basins and the water channel, at a distance of 173.45 m. to the west of the eastern

9. One would have to assume that the Hellenistic starting line was installed ca. 270 B.C. without the *husplex* and that at some later date within the third century B.C. the *husplex* was added.

10. Charles K. Williams, II and Joan E. Fisher, "Corinth, 1970: Forum Area," *Hesperia* 40 (1971): 22.

11. Several inscribed bases have been found in the area of the racecourse honoring Isthmian victors, I-790; I 69 3. See Charles K. Williams, II, "Corinth, 1969: Forum Area," *Hesperia* 39 (1970): 38-39. See also Charles K. Williams, II, "Pre-Roman Cults in the Area of the Forum of Ancient Corinth," dissertation in Classical Archaeology, University of Pennsylvania (1978): 24-25.

12. The quadriga base faced east so that the runners leaving the starting line at the east end of the *dromos* would have run toward the sculptural group, passing it to their left.

13. See Charles K. Williams, II and Orestes H. Zervos, "Excavations at Corinth, 1989: The Temenos of Temple E," *Hesperia* 59 (1990): 351-356.

14. The cuttings may be the lines of the foundations of a stoa. See Williams (1980) 28, note 42. See also Robert L. Scranton, *Corinth* I, iii, *Monuments in the Lower Agora and North of the Archaic Temple*, Princeton (1951): 76-77.

15. Charles K. Williams, II, "Pre-Roman Cults in the Area of the Forum of Ancient Corinth", dissertation in Classical Archaeology, University of Pennsylvania (1978): 142.

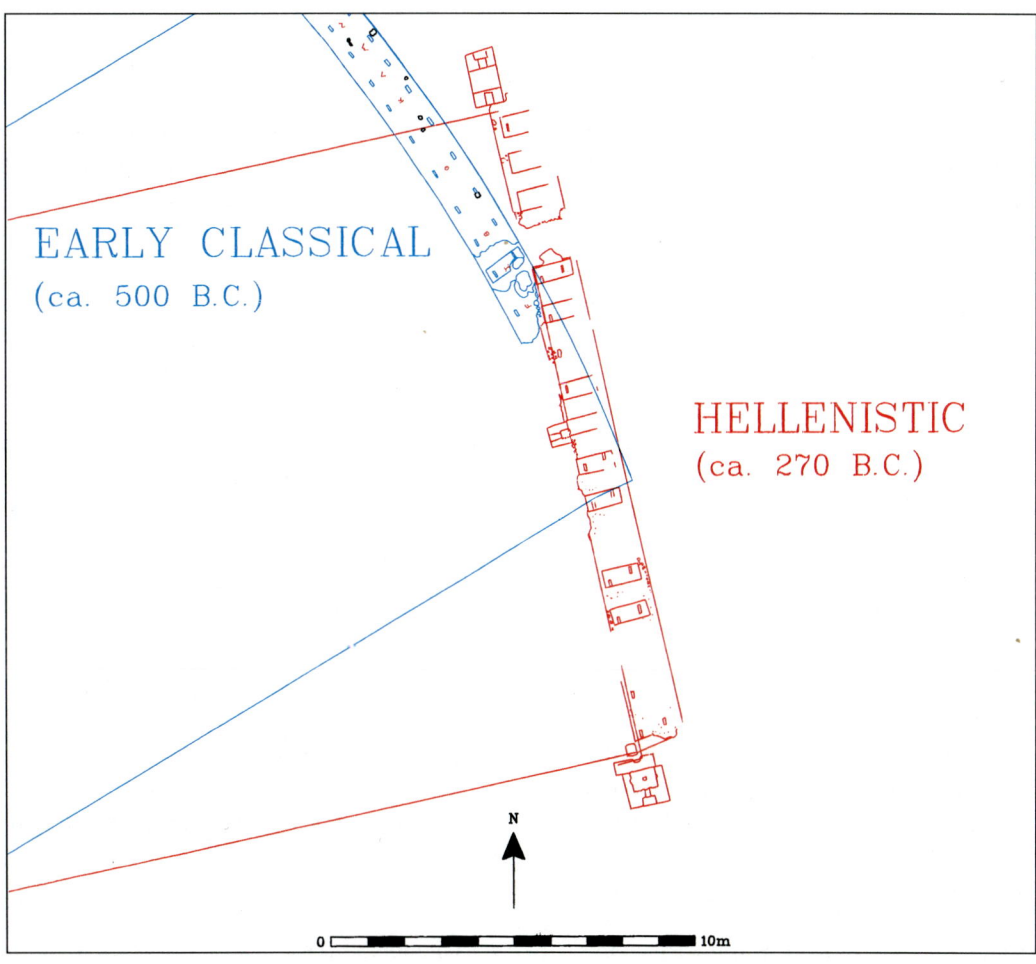

EARLY CLASSICAL
(ca. 500 B.C.)

HELLENISTIC
(ca. 270 B.C.)

N

0 ⊏━━━━━━━━━━━━━ 10m

45. Corinth, Detail of Early Classical and Hellenistic starting lines.

starting line.[16] Another water system is known on the south side of the eastern end of the Hellenistic *dromos*, both at the starting line and around the north side of the curved terrace (Fig. 24). The water channel is cut into the upper surface of the curved terrace wall and empties into three water basins. The channel runs along the full length of the eastern side of the starting line, bending toward the northwest at its northern end. The addition of hydraulic facilities in the Hellenistic *dromos* form an important change from the Archaic *dromos* which had no known water facilities. Water basins of the type that are found along the south long side of the *dromos* are at track level and are likely to have been used in connection with the maintenance of the racecourse surface. The crushed-poros levels of the track surface would become hardened by the sun, as well as by use, and it would have been necessary to sprinkle water periodically on the track surface to make it more resilient for the athletes.[17] The water basins that are found along the north side of the curved terrace are elevated above the level of the racecourse floor, closer to the level of the curved terrace floor surface and could have been used for varied purposes.

It is clear from the evidence of the starting line of the Hellenistic *dromos* that the athletic events of the Hellenistic period differed from those of the Archaic period.[18] The straight Hellenistic starting line suggests very clearly that races were run in parallel lanes and that these races were probably one length of the *dromos* (Figs. 46, 47). The single length race is suggested because of the width of the lanes (1.03 m.) which would not allow enough space to turn adequately at the west end of the *dromos* (around individual turning posts) without running into neighboring lanes.[19] Also, the mechanical starting device (*husplex*) suggests a short race in which the start was critical. Although a detailed and accurate reconstruction of the *husplex* may be impossible at this time, the structure is likely to have been composed of three vertical elements, one at each end of the starting line supported by stone foundations and one in the center against the west edge of the line. A horizontal element, either a wooden bar or a cord, would have been positioned in front of the starting athletes and would have been raised or lowered abruptly for the start.

Either the events of a particular athletic festival at Corinth had changed over time or it became advantageous to run the short distance races from east to west in the realigned racecourse of 270 B.C. Before that time races would have been run west to east. Because of the similarity in the physical design of the Hellenistic and the early classical starting lines, it is even possible that the component parts of the straight starting line could have been moved from the west end of

---

16. This water basin shares some similarities with the reservoir basin at the northeast corner of the stadium at Epidauros which is located over 15 meters behind the starting line. It must have served as an intake basin for the entry of water into the stadium water channels from the east. See Romano, "Stadia" (1981): 17-18.

17. The usual interpretation of water channels and water basins at ground level in the context of the Greek stadium or gymnasium is that they were for the athletes, as well as for the spectators, when appropriate, to drink from. But this is unlikely to have been the case for a number of reasons, primarily sanitation. It is more likely that these water facilities in athletic contexts were related to track maintenance and care. See Romano, "Stadia" (1981): 215-217.

18. It must be admitted, however, that the design of the poros starting line blocks with widely spaced toe grooves is very similar to the blocks of the Archaic *dromos*. The widely spaced position, in itself, probably does not indicate the variety of footrace to be run.

19. A possible solution would have been to start runners from every second lane, thus creating a blank lane between each starting runner.

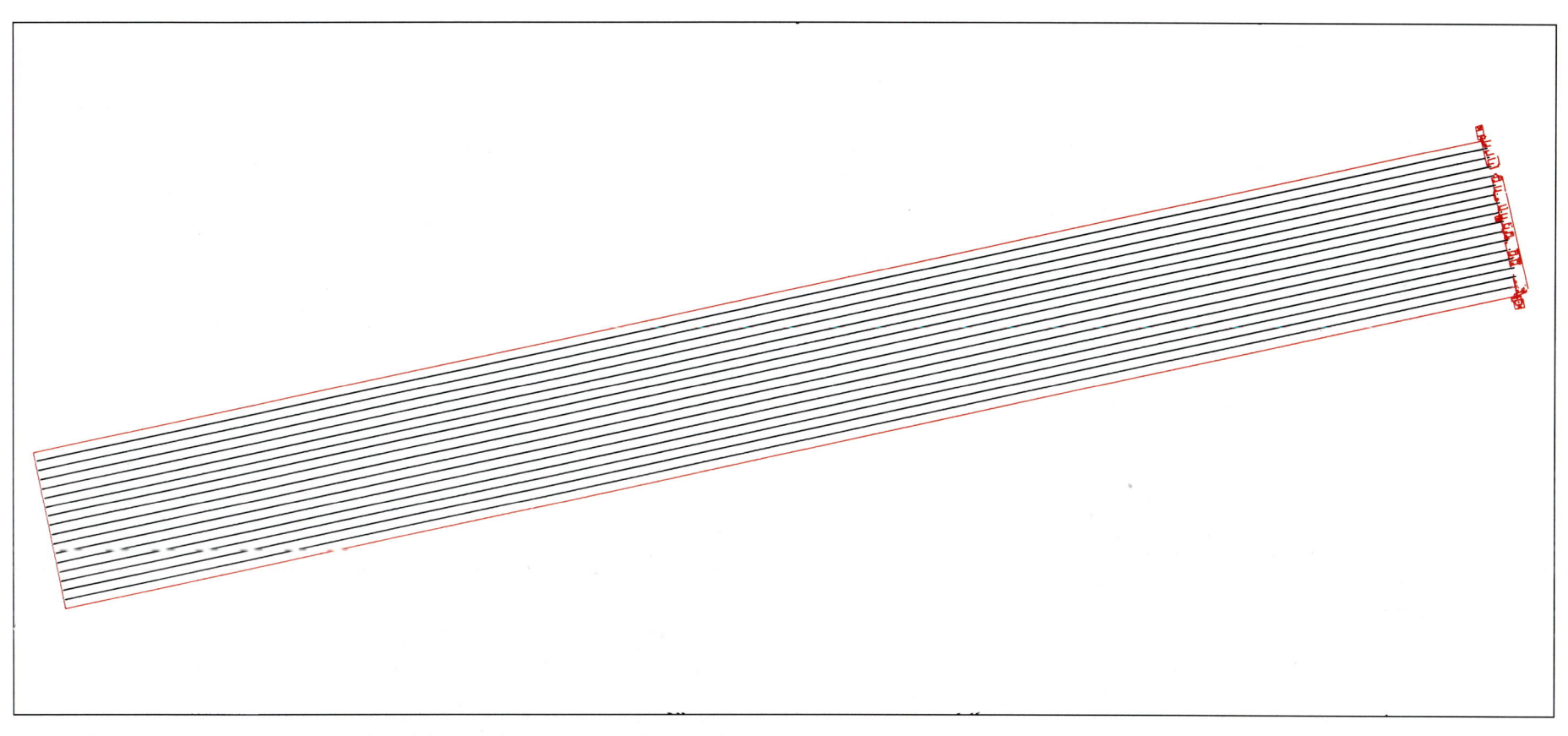

46. Hellenistic starting line and restored parallel lanes of racecourse.

the Early Classical racecourse, to the east end of the Hellenistic race-course.[20] This line could have provided the athletes in the Early Classical period a starting line for the short distance footraces, *sta-dion*, *diaulos*, and possibly *hippios dromos*, which were run from west to east.

The successive starting lines at Corinth, characterized by individual toe grooves for the starting athletes, are clearly unlike the two parallel groove starting lines that have been found at a number of other Panhellenic sanctuaries, Delphi, Olympia, Nemea, Epidauros, Isthmia as well as at the local sanctuary at Halieis.[21] Although the parallel groove starting line types are difficult to date, it seems most likely that they were first introduced in the Hellenistic period and then used through the Roman period.[22] It should not be surprising, therefore, when considering the Early Classical starting line in Corinth that there would have been a different type of design. The earliest starting system, *balbides,* at the nearby Sanctuary of Poseidon at Isthmia dates generally to the fifth century B.C. and has no parallel in the Greek world.[23] However, the *balbides* sill at Isthmia is clearly that for a short race, a *stadion* or *diaulos* since the athletes are running in parallel lanes. Since the eastern end of the Earlier Stadium at Isthmia

has not been excavated, we do not know what type of starting line was in use there.[24] The Corinthians may well have been experiment-ing with a number of different starting systems at this time.[25]

There are few well known depictions of runners at the start of a footrace that depict such a wide-spaced starting position, as indicat-ed by the Early Classical and Hellenistic starting grooves at Corinth.[26] There is also an interesting floor mosaic from the Agonotheteion in the South Stoa at Corinth dated to the second half of the first century A.D. The mosaic depicts a nude athlete holding a palm branch stand-ing before a seated female figure, the Goddess of Good Fortune. The athlete, who is likely to be a runner, has his feet widely spaced apart, left foot forward, and may have the toes of his feet in starting grooves, a feature which has been previously interpreted as shadows (Fig. 48). It is possible that this mosaic may have been based on an earlier paint-ed representation of a victorious athlete in Corinth, since the Hellenistic *dromos* and starting line would have been covered and no longer visible by the second half of the first century A.D.

In addition, there have been found from the large rectangu-lar peristyle courtyard at the Villa of the Papyri in Herculaneum, two nearly identical bronze athletes that have been variously described as

20. I thank Benjamin Schoenbrun for this suggestion.

21. However, as mentioned above in Chapter 1, p. 28, note 60, at Nemea there are two individual blocks that have similar toe holds to those found in Corinth, suggesting that the individual toe holds are a predecessor to the con-tinuous groove blocks in the evolution of starting line design.

22. See Romano, "Stadia" (1981): 205-216.

23. See above Chapter 4 p. 83 note 4.

24. Broneer suggested that the *stadion* length of the Earlier Stadium at Isthmia was originally 192 m. (the same length as the Olympia III Stadium) and then reduced to 181 m. after the introduction of the single groove starting line (Broneer, *Isthmia,* Volume I, *Temple of Poseidon,* Princeton (1971): 174-181.) Elizabeth R. Gebhard has followed this line of thinking in "The Sanctuary of Poseidon on the Isthmus of Corinth and The Isthmian Games," in *Mind and Body* (1989): 82-88, and in Gebhard and Hemans (above Chapter 1, p. 26, note 53,) p. 59, note 135. She postulates a necessity for bringing in at

least 5,000 cubic meters of soil in order to raise the east end of the *dromos* of the Earlier Stadium 5.30 m. to provide a level racecourse surface. I suggest that the major earth moving project would not be necessary if the length of the Earlier Stadium at Isthmia was the same as the *dromos* in Corinth, ca. 165 m. I would assume that the same foot measure was used by the Corinthians to lay out the two closely contemporary racecourses just as it appears that the same foot was employed in measuring the (stylobate) length of the Archaic Temple of Apollo at Corinth and the Classical Temple of Poseidon at Isthmia. See Broneer, *op.cit.,* pp. 177-181, and Romano, "Stadia" (1981): 250-267.

25. See Broneer, *Isthmia* II (1973), Appendix II, pp. 137-142.

26. See above Chapter 2, p. 67, note 45. Two marble statues from Velletri, now in the Palazzo dei Conservatori, Rome have been identified as runners, although their right legs are in advance of their left. See Walter Woodburn Hyde, *Olympic Victor Monuments and Greek Athletic Art,* Washington, D.C. (1921): 198. See Broneer, *South Stoa* (1954): 107-109; plates 30-31.

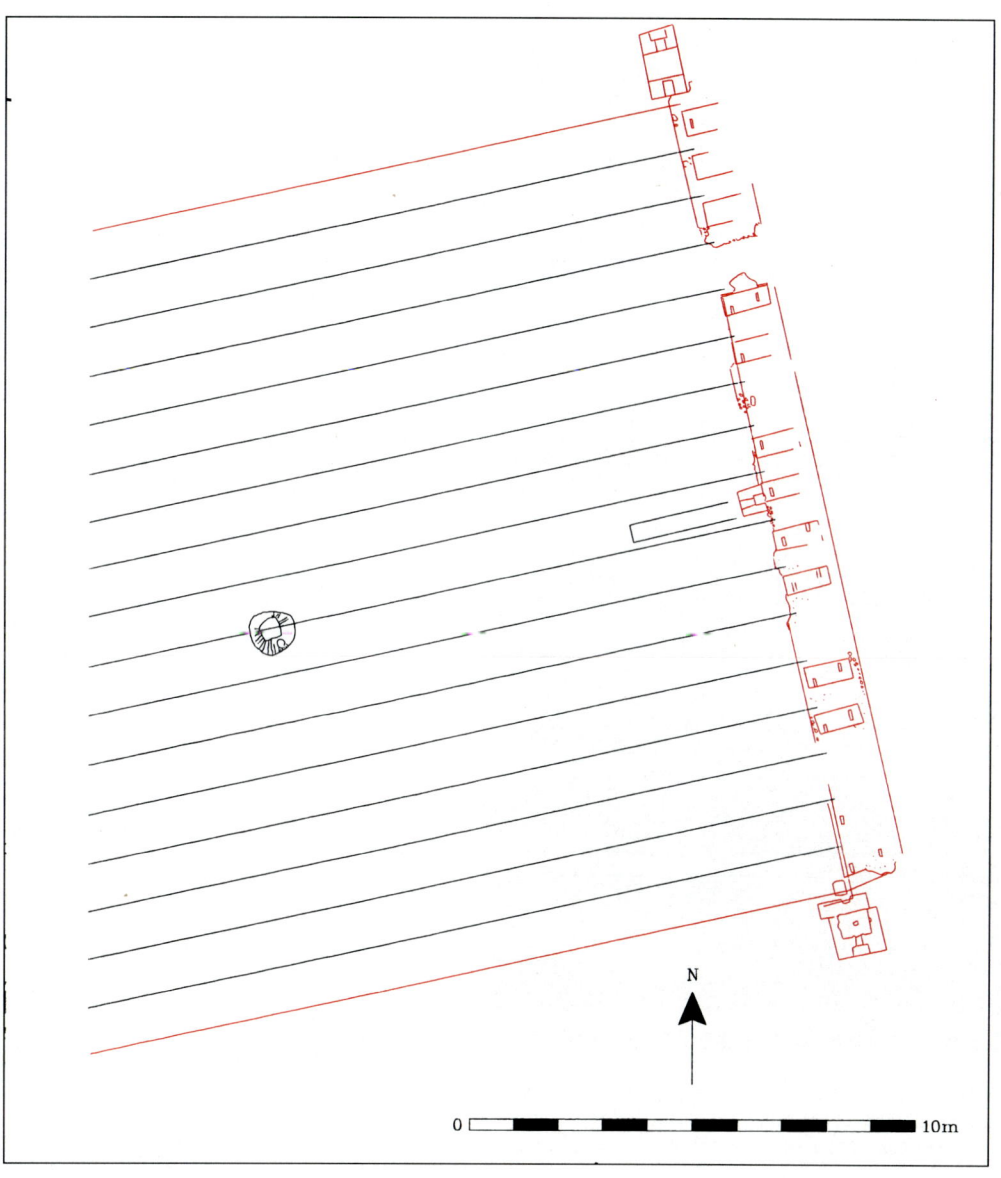

47. Detail of Hellenistic starting line and parallel lanes, including large post hole in racecourse surface.

48. Roman floor mosaic from the "Agonotheteion" of the South Stoa in Corinth, second half of the first century A.D. The central panel depicts an athlete standing before the goddess of Good Fortune. Courtesy of the American School of Classical Studies at Athens, Corinth Excavations.

wrestlers or runners.[27] These athletes are more likely to be standing at the start of a footrace, and their foot positions are compatible with the requirements of the Corinthian starting lines (Figs. 49a, 49b). The athletes were originally located flanking the long sides of the long and narrow pool, which may have been a metaphorical representation of a racecourse.

27. A discussion of the Herculaneum athletes as runners at the start of a foot race in the context of the sculptural program of the villa, is considered in a forthcoming article in *Art History* by P. Gregory Warden and David Gilman Romano. "The Course of Glory: Greek Art in a Roman context at the Villa of the Papyri at Herculaneum." The bronzes are in the National Archaeological Museum, Naples, #5626 and 5627. They are often believed to be copies of late fourth century B.C. Greek originals. See Maria Rita Wojcik, *La Villa dei Papiri ad Ercolano*, Rome (1986): 108-111; Richard Neudecker, *Die Skulturenaustattung römischer Villen in Italien*, Mainz am Rhein (1988): 154.

49a. Bronze athletes from Villa of the Papyri, Herculaneum, identified here as starting runners. Naples Museum # 5626. Photo courtesy of the Deutsches Archäologisches Institut, Rome.

49b. Naples Museum # 5627.

Chapter 6
GREEK DESIGN ELEMENTS OF THE ROMAN CIRCUS

The Roman circus was a totally separate and distinct architectural entity from the *stadion* and the *dromos*. The circus had as its predecessor the Greek *hippodrome* and both the Greek *hippodrome* and the Roman circus were used for equestrian contests as well as for the accommodation of spectators.[1]

The architectural development of the circus has been considered and there is some evidence for its evolution. It is known, of course, that equestrian contests formed an important aspect of Greek athletics from earliest recorded times. From the *Iliad,* Book 23, is the description of the Funeral Games of Patroklos where greatest importance is given to the chariot race. Although Greek chariot racing was an important element of Greek athletic festivals there is very little surviving architectural evidence for the structure where it took place.[2] In fact, there are no excavated Greek *hippodrome*s. The best preserved ancient *hippodrome* from mainland Greece, is located near the summit of Mt. Lykaion in Arkadia approximately 1200 m. above sea level, at the sanctuary of Zeus Lykaios, the site of the Lykaion Games.[3]

From literary and historical sources, we do have some information about other *hippodrome*s in mainland Greece. According to traditional accounts the first chariot race, the *tethrippon,* was added to the 25th Olympic Games in 680 B.C. The first horseback race,

keles, was introduced in 648 B.C. and the first two horse chariot races, *sunoris,* in 408 B.C. The *hippodrome* at Olympia is known to us from the account of Pausanias (VI,19,10 - VI,21,1).[4]

It is also known that the Etruscans practiced chariot racing from at least the sixth century B.C. although there are no discovered examples of an Etruscan *hippodrome*. We also lack substantial historical and literary evidence about them. Most of the extant evidence about Etruscan contests comes in the form of tomb paintings and sculptural representations. There is, for instance, the painted representation of the Tombe della Bighe from the second half of the sixth century B.C., where there are painted representations of the chariots as well as spectators and seating areas.[5]

About the early development of the Roman circus we have good information. Circuses were popular throughout the Roman Empire and were commonly found in Roman cities and towns. There are numerous Roman circuses that have been excavated and studied and the theory and practice of the typical starting arrangements are fairly well known.

For example, there is considerable information about the history and evolution of the Circus Maximus in Rome. Traditional sources credit the Elder Tarquin, ca. 600 B.C., with the earliest construction of the circus.[6] Of course, it is not known what the specific

---

1. The most comprehensive work on the Roman circus is the recent book by John H. Humphrey, *Roman Circuses, Arenas For Chariot Racing,* Berkeley, 1986.

2. Spectators and spectator facilities rarely appear in painted scenes of athletic contests. There are only three examples, to my knowledge, of vase painting scenes: the well known fragment of a *dinos* by Sophilos, Athens National Museum 15499, ABV 39.16; the fragment of a neck amphora in Berlin of the Tyrrhenian Group, Berlin 1711, *ABV* 95.8; and an Attic Black Figure of Panathenaic shape, Bibliothèque Nationale 243. All three of these painted scenes involve chariots or horses and the spectator facilities must, therefore,

be interpreted as those of hippodromes as opposed to stadia.

3. Romano, "Stadia" (1981): 172-177. From the archaeological evidence as well as from the description of Pausanias (VIII,38,5), it seems likely that the hippodrome was constructed around the stadium.

4. Humphrey (1986): 7, note 9.

5. See F. Weege, "Etruskische Gräber mit Gemälden in Corneto," *JdI* 31 (1916): 105-168, Supplement 1. See also Massimo Pallottino, *Etruscan Painting, Lausanne* (1952): 61-64.

6. Livy (1,35,8).

architectural arrangements were in the Circus Maximus at this early stage. Livy states that the *carceres*, the starting compartments, were first built in the Circus Maximus in 329 B.C.[7] although it is not clear what the starting arrangements were before this time. There are, unfortunately, no details about the fourth century starting gates.

Although there is no parallel for the curved design of the Early Classical starting line of the *dromos* from Corinth from an excavated stadium in the Greek world, there are good parallels for the general design of the starting line from known examples of the Roman circus.[8]

Roman circuses commonly have curved starting lines and compartments, *carceres,* for the chariots at one end of the racecourse. The arc of the starting line is taken from a circle with a very large radius, the center of which is a distant point on the racecourse floor. In the circus at Lepcis Magna, datable to the second half of the second century A.D., the center of the circle is a point near the south wall of the circus some 300 meters to the east (Fig. 50). This point is also located near the midpoint of the projected length of the *spina* and opposite the tribunal box. The reason for a circle with such a large radius is related to the width of the chariots and the orientation of the arc of the *carceres*. A larger circle provides a better orientation for the chari-

ot teams down the long axis of the circus. This point may also have been on or near the finish line.[9] The average width of each of the starting positions at Lepcis Magna is equal to approximately one degree (50' 36") of the circle.[10] In practice the chariots would run in lanes to the right of the first meta, turning post, providing each of the chariots enough width to pass them without a collision (Fig. 51). At the first *meta* there would commonly be a white line, *alba linea,* drawn on the racecourse floor which would serve as a break line for the chariot teams. There is considerable archaeological and literary evidence for the presence of white lines in the Roman circus. Humphrey summarizes this evidence and discusses three different types of white lines. The first variety were the white lines that demarcated the lanes for the chariot teams in the opening stage of the race. These white lines likely connected the starting gates with the break line. The second variety was the break line itself, which was the point at which the chariots could leave their assigned lane designations and head for the inside position. The third variety of white line was the finishing line, which may have been in some cases an extension of the break line, but which more commonly was a line approximately half of the way down the racecourse on the right side.[11] This break line was created by drawing a concentric circle from the same center point near the south side of

---

7. Livy (8,20,2)

8. A curved line of stone blocks, within the racecourse floor of the Olympia III Stadium at Olympia has been tentatively identified as a starting line dating generally to the Roman period. See above Chapter 1, p. 23. It has been suggested by the excavators that this starting line may be associated with the footraces in honor of Hera that are described by Pausanias (V, 16, 2-3). See Alfred Mallwitz, "Das Stadion," *OlBer* VIII, Berlin (1967): 52-53. This line of blocks is located approximately 153 meters to the east of the eastern starting line. A preliminary computer study has shown that the circle, of which

the starting line forms an arc, has a radius of ca. 120-125 m. See also Romano, "Arete" (1983): 9-16.

9. See Humphrey (1986): 21, fig. 7.

10. This figure is based on a preliminary study of a drawing of the starting assembly only. I do not have survey data available comparable to that of the Early Classical starting line in Corinth. A circle with positions of one degree is within the range of possible circles.

11. Humphrey (1986): 84-91.

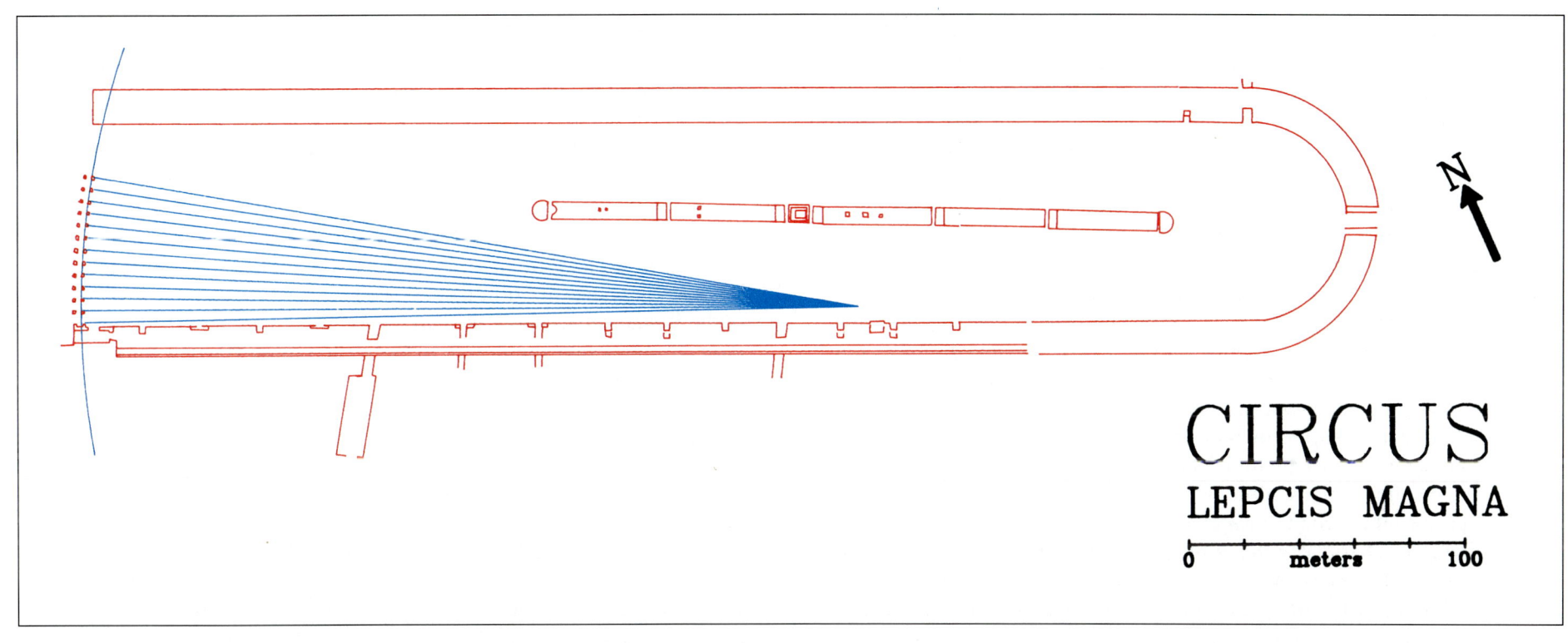

CIRCUS
LEPCIS MAGNA

0   meters   100

50. Drawing of circus of Lepcis Magna showing the focal point of the starting positions (blue lines). Drawing after John Humphrey, *Roman Circuses.*

the circus as was used to create the arc of the starting gates. In theory, all of the chariot teams had an equal distance to run from the *carceres* to this break line in lanes and had a chance of arriving at this early stage of the race at the same instant (Fig. 52).[12] It is also common in the Roman circus for the two *metae* to be joined by a *spina,* a low and continuous barrier wall, although this structure is usually not exactly parallel to the axis of the racecourse. Oftentimes the *spina* is canted slightly so that the space to the right of the near, *meta secunda,* where the chariots would first pass the break line, is wider than the area to the left of it.[13] The *spina* is commonly equidistant from the borders of the racecourse at the far *meta prima.* The angle of the *spina* probably compensated to some degree for the slightly unequal distances run from the *carceres* to the break line, especially if the break line was straight and not curved, as it appears in the Lyon circus mosaic.[14] In addition, the chariots on the outside of the racecourse would have a longer distance to run to the *meta prima* than would the chariots on the inside positions, although chariots to the outside would likely approach the distant *meta* at a more advantageous angle for a safe turn. The chariots on the inside positions would likely have felt the squeeze of approaching chariots from the right as the race approached the far *meta* (Figure 51). Roman chariot races were commonly run over a distance of seven laps.

There appears to be some similarity between the design of the early starting line at Corinth and the overall design of the Roman cir-

cus. Humphrey has written "Roman *circus*es generally have regular and predictable designs."[15] This is very true, although the earliest origin of the design of the Roman circus has until now not been at all clear. The "canonical" Roman circus dates generally to the time of Imperial Rome.[16] Difficulty in understanding the origins of the Roman circus has come from the lack of excavated Greek *hippodrome*s, the architectural predecessor of the Roman circus.

The observation that the geometric design principles of the "canonical" Roman circus is so similar to the design of the starting line and racecourse in Corinth of ca. 500 B.C. suggests that the design of the Roman circus may have had its origin in an earlier Greek *stadion.* It seems that the design of both the Greek starting line in Corinth and the "canonical" Roman circus would have been based on a number of simple geometric principles, that would attempt to provide a fair and equal start to the athletes or chariots. The other possibility is that the Greek architects in the sixth and fifth centuries B.C. and the Roman architects in the second century A.D., with similar design necessities, actually came up with very similar solutions to the construction of starting lines and starting gates. This possibility seems less likely. In modern track and field it is common to find a curved starting line for a distance race of 1500 m. or greater. Of course, the nature and design of the modern track is totally different and of independent origin than the ancient, but the curved starting line serves a similar purpose in each: to provide an equal distance to be run from

12. Humphrey (1986): 21, writes that "in the canonical Roman circus, the optimum figure for the distance from the gates to the inside of the barrier seems to have been estimated at about 140-160 m., which represents about one-third the length of the entire arena."

13. Such is the situation in the circus of Lepcis Magna where the space between the *meta secunda* and the south limit of the racecourse floor of the circus is 36.9 meters and the distance to the north is between 27-28 meters. The spaces are equalized at the *meta prima,* approximately 32 meters for each. See Humphrey (1986): 23, fig. 6.

14. Although the *carceres* appear to be straight in the same mosaic. See Humphrey (1986): 84-91.

15. Humphrey (1986): 60.

16. Humphrey (1986): 23, states that "while the form of the canonical circus is fairly clear, it is much more difficult to date the moment or moments when the refinements were first introduced. One may suspect that they originated in the Circus Maximus and certainly seem to be present in the circus by the time of Trajan's reconstruction."

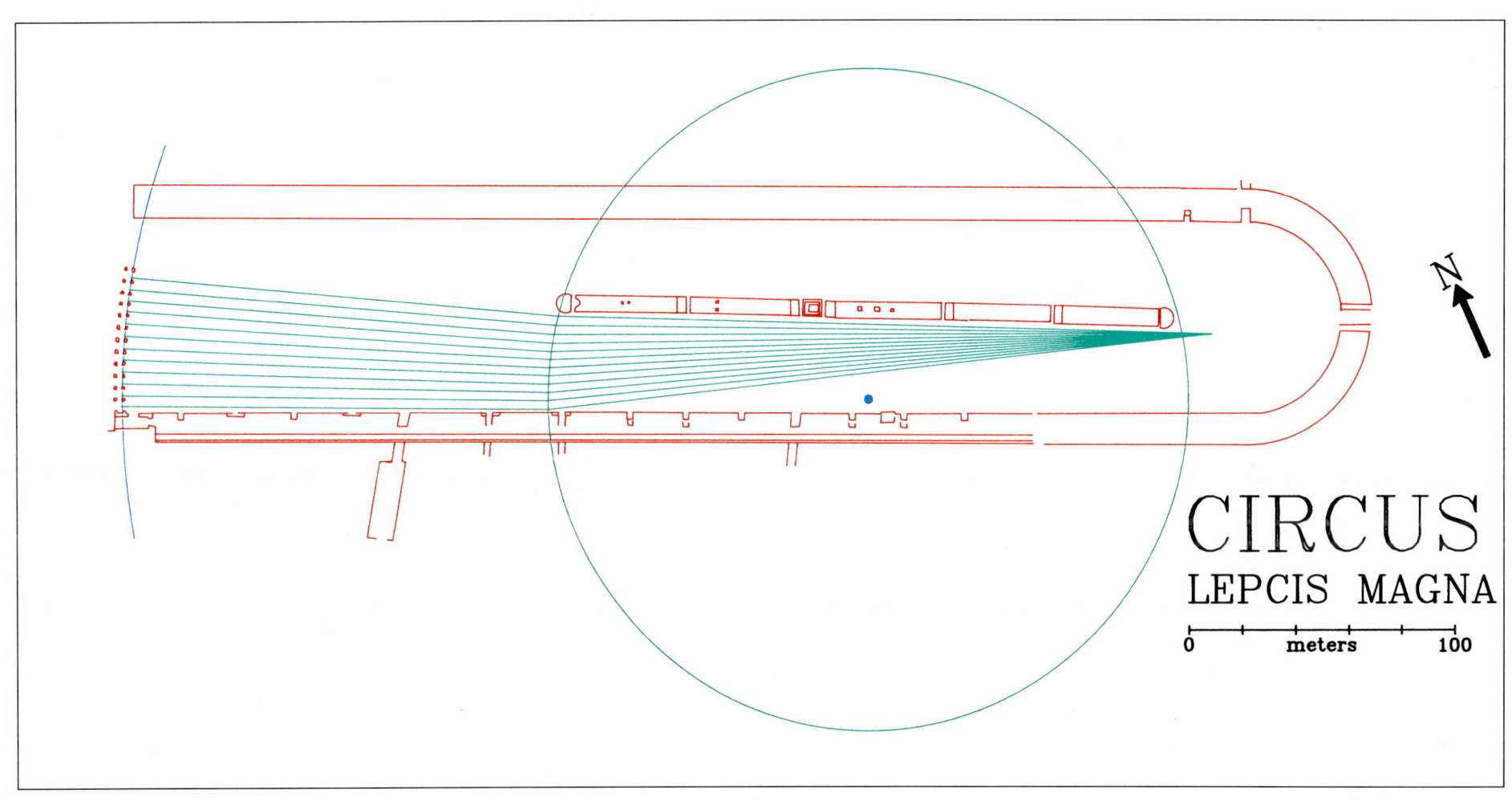

CIRCUS
LEPCIS MAGNA

51. Drawing of circus of Lepcis Magna showing method of creating "break line" and paths (in green) to be run by individual chariots.

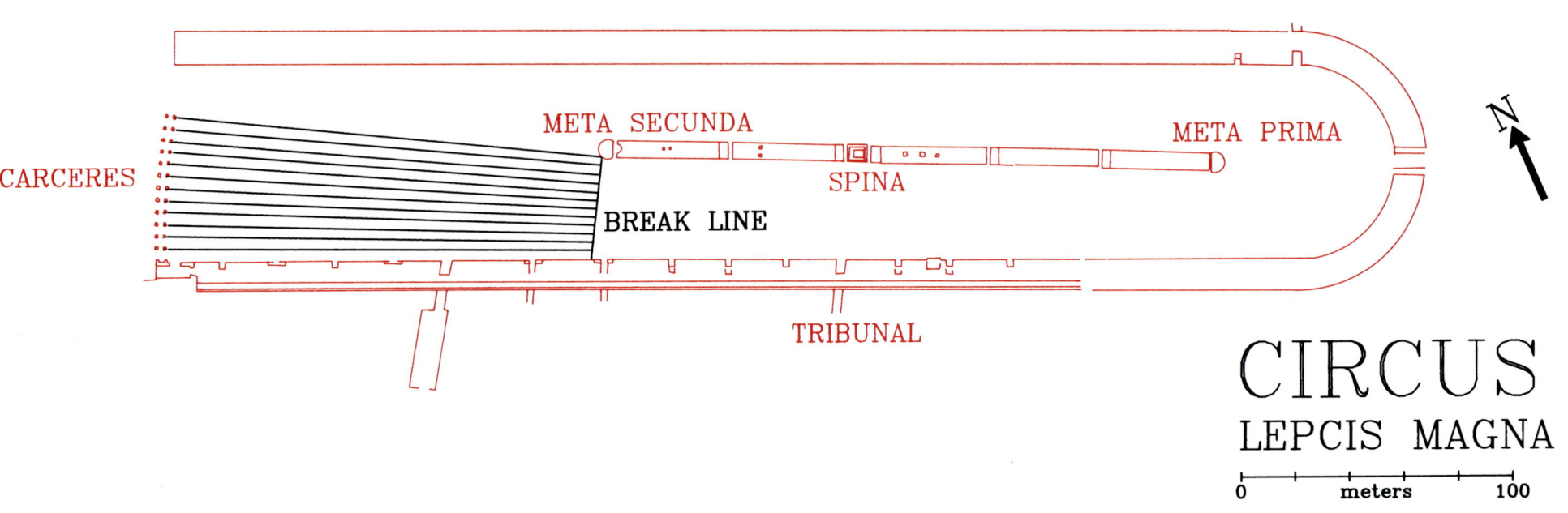

CARCERES

META SECUNDA

META PRIMA

SPINA

BREAK LINE

N

TRIBUNAL

CIRCUS

LEPCIS MAGNA

0   meters   100

52. Drawing of circus of Lepcis Magna illustrating *carceres* and lanes to be run in by the chariots to the "break line."

starting positions to a break point (Fig. 53) Modern track and field engineers lay out a curved starting line with a system very similar to the ancient system.[17] In the modern day a point on the tangent of the curve of the track is chosen, one foot outside the curb of the inside lane, as an equidistant point from all starting positions. At Franklin Field the lanes are 42 inches wide.[18] Using a cord, an arc is swung from the equidistant point across the track to create the starting curve.

How, specifically, had the Romans by the time of the construction of the circus at Lepcis Magna, e.g., modified the concept of the earlier design of the Classical *dromos* at Corinth? The Romans extended the focal point of the starting positions well past the near *meta secunda* to nearly two-thirds the length of the course. The break line in the Roman circus was created by drawing a concentric circle from the same focal point as opposed to a tangent circle in the Greek system. Two features of the design remained the same. The break line remained at approximately one third length of the racecourse and the width of the individual starting positions remained at approximately one degree of the circle.

53. Photograph of the start of the 3000 m steeplechase from the Penn Relays, 1982, Franklin Field, University of Pennsylvania, Philadelphia, where starting athletes are using a curved starting line. Photo by the author.

17. I thank Don W. Handley, professional engineering consultant to the Department of Athletics at the University of Pennsylvania, for discussing with me these points at Franklin Field, October 18, 1991. I also discussed the same questions in 1980 with Mr. Irv. Mondshein, then track coach at the University of Pennsylvania.

18. This is the equivalent of 1.07 m. which is close to the widths of the parallel lanes from the Hellenistic *dromos* in Corinth (1.03 m.) as well as the parallel lanes from the fifth century *balbides* sill at Isthmia (1.05 m.).

Chapter 7
CONCLUSIONS

There is substantial literary, historical and archaeological evidence to suggest that the formal appearance of the *stadion* as a structure as well as a footrace may be dated at least as early as the mid-sixth century B.C. Although the specific details remain obscure, it may have been at some time during the seventh or the first half of the sixth century B.C. that low sloping artificial embankments were first introduced as a means of providing accommodation for increasing numbers of spectators at athletic contests. Although it is not known exactly where this happened, it may have been at Olympia, traditionally the oldest and the best known of the Panhellenic athletic festival sites. It is known, for instance, that Olympia experienced a dramatic increase in the number of visitors during the early seventh century B.C. as can be documented by the increased number of wells to the east of the *altis*.[1]

The word *stadion*, in its earliest use probably meant "standing place," while later it came to be associated with the length of the *dromos* and the word for the footrace. It may be significant that it is the embankments that appear chronologically early, ca. 540 B.C. at Olympia, and possibly 580 B.C. at Isthmia, and the word *andronstadio* appears ca. 550 B.C. on the Panathenaic amphora in the Metropolitan Museum of Art (1978.11.13). Only later, in the fifth century B.C., is the word *stadion* first defined as a linear measurement by Herodotus.

On the other hand the word *dromos* has a longer history. As a simple racecourse, a level stretch of ground, the word is known from Homeric works where there is no specific length given for the racecourse nor are there any formal facilities for spectators of any kind.

If the *Iliad* and the *Odyssey* were written down in the late eighth century, as is commonly believed, there exists a kind of *terminus post quem* for the introduction of the word *stadion* (since Homer doesn't use it). The dates of ca. 550 B.C. for the Metropolitan Museum Panathenaic amphora and 540 B.C. for the first artificial embankment at Olympia (possibly 580 B.C. for the first artificial embankment at Isthmia) and ca. 550 B.C. for the "Gorgos" seat block at Olympia, as a kind of *terminus ante quem*. It seems likely that at some time between the late eighth century and ca. 600 - 550 B.C. embankments were introduced for the growing numbers of spectators at athletic festival sites and, as a result, the other two meanings of the word came into existence by association with the primary one.

Of course there must have been a *dromos* at Olympia earlier than the ca. 540 B.C. artificial embankment although, unfortunately, nothing of it remains.[2] It seems likely that the natural slope of the Kronos hill to the north of the earliest *dromos* of Stadium I was the original embankment, possibly the earliest "standing place" as *stadion* from the seventh century B.C.

If the *Olympic Register* is correct, there must have been a racecourse, *dromos*, at Olympia since the eighth century B.C.[3] An open question remains whether the earliest *dromos* at Olympia, as racecourse only, was 600 feet long? Another question concerns the

---

1. See above Chapter 1, pp. 17-24.
2. See Alfred Mallwitz, "Cult and Competition Locations at Olympia," in Raschke (1988): 98-99. See above Chapter 1, pp. 17-19.

3. Mallwitz argues that the athletic contests did not begin until about 700 but that the cult is more ancient as demonstrated by Protogeometric finds. He argues for a one-year cycle between early Olympiads. See Alfred Mallwitz, "Cult and Competition Locations at Olympia," in Raschke (1988): 99-101.

length of the foot and whether it was the same foot used throughout the history of the festival. When was the specific distance of 600 feet chosen as the length of the footrace at Olympia or elsewhere? Although these questions cannot be settled at the present, there is some evidence that suggests that the specific 600 foot length may have been a borrowing from Babylonia, possibly the product of the basis of the two Babylonian mathematical systems, 60 x 10. This borrowing would have most likely occurred at the time of the heaviest influence from the eastern part of the world, known in the Orientalizing Period, roughly the late eighth and the seventh century B.C. It is even possible that some of the athletic events themselves may have come as a part of the same eastern influence around the same time.[4]

Important developments of the Greek Orientalizing Period occurred in Corinth, for instance the introduction of the Doric Order to Greek architecture,[5] the introduction of the terracotta roof tile, painted metope decoration for temples, as well as Black-figure vase painting and orientalizing motifs on painted pottery. It may have been, therefore, that there was at Corinth a 600 foot *dromos* from as early as the seventh century, although the excavated example from the Upper Lechaion Road Valley can only be dated to the sixth century B.C. The knowledge and use of a numerical equivalent of π at Corinth, ca. 500 B.C., must also be due to eastern influence, likely Egyptian or Babylonian.

The essential elements of the *stadion* included *dromos* and embankments, either natural or artificial, in the context of the sanctuary. We know the most about the design and function of the Archaic

dromos in Corinth. In the years around 500 B.C. an architect with the knowledge of plane geometry including the properties of circles and with the design requirements of a distance footrace in mind set out a *dromos* of a *stadion* in length in Corinth. At the east end of the racecourse he designed a starting line for a long distance footrace for seventeen athletes. The architect's design attempted to provide for an equally fair start for each athlete, assuring the same distance to be run toward a distant point. The starting line provided individual toe grooves for the feet of each athlete which were widely although unequally spaced apart. The reason for the wide spacing may have been to provide the fairest start possible. The starting line was characterized by no other posts as lane dividers or turning posts, and there is no evidence for a mechanical starting mechanism, as the *husplex* in the Hellenistic starting line. The painted letters as numbers on the surface of the Early Classical starting line demonstrate clearly that the starting line was used for competition, although it is not known with what cult or cults the racecourse was associated. Both the early and the late racecourses are situated adjacent to the Sacred Spring; the early racecourse was immediately adjacent to the Heroon of the Crossroads.

In the years after 270 B.C. the Archaic *dromos* was replaced by the Hellenistic *dromos*. Although certain modifications were made to the nature of the racecourse, including the addition of hydraulic facilities, and a straight starting line at the east end, many of the essential elements of the Early Classical starting line design were retained.

Spectators could have been accommodated to the east of the starting line, on the rising ground to the south, as well as on some

4. See above Chapter 1, pp. 7-13.
5. J.J. Coulton thinks that much of the earliest stone architecture came from Greek familiarity with Egyptian monuments. See J.J. Coulton, *Ancient Greek Architects at Work,* Ithaca (1977), chapter 2. Charles K. Williams, II, in a

review of Coulton suggests that heavy wooden superstructure came in suddenly in the seventh century B.C. in order to support the terracotta roof and therefore the beginning of the Doric form. See *The Art Bulletin* 62 (1980): 151-153.

parts of the south slope of Temple Hill. These areas could have afforded an excellent view for a moderate number of spectators. Although it is not known how often the athletic festival was, or festivals were, held in this location, the fact that successive racecourses were used in the area between ca. 500-146 B.C. suggests that one of the principal functions of this area of town may have been as a festival ground.

Since so little of the extent of the early racecourse is known from excavation, there may have been important features of the original facility that are now destroyed. It would not be surprising if a starting line of very different design were once present at the west end of the course. For it is likely that footraces of other lengths would have been run on the *dromos*, short races, such as the *stadion* and *diaulos*, necessitating a starting line of a different design.

It is, of course, impossible to know if the unique curved Early Classical starting line and the design of the Archaic *dromos* at Corinth was the first of its kind in the Greek world. Certainly it is an early form of the Greek *stadion* as a 600 foot *dromos* and the oldest example of a starting line designed specifically for a long distance race. It also demonstrates the earliest use of π in the Greek world and clearly attests that the Greeks had knowledge by ca. 500 B.C. that a circle was composed of 360 degrees. It is very likely that a similar type of starting line was employed in contemporary and later examples of the Greek *hippodrome* and thereafter the same principles were used and applied to the design of the Roman circus.

BIBLIOGRAPHY

Altenmüller, Hartwig, and Ahmed M. Moussa. "Die Inschriften auf der Taharkastele von der Dahschurstrasse," *Studien zur Altägyptischen Kultur* 9 (1981): 57-84.

Aupert, Pierre. *Fouilles de Delphes, II, Topographie et Architecture,* part 7, "Le Stade," Paris, 1979.

_____. "Le cadre des jeux Pythiques," *Proceedings of an International Symposium on the Olympic Games (1988),* William Coulson and Helmut Kyrieleis, eds., Athens (1992): 67-71.

Beazley, J.D. *Attic Black-Figure Vase-Painters,* Oxford, 1956.

_____. *The Development of Attic Black Figure,* revised edition, ed. by Dietrich von Bothmer and Mary B. Moore, Berkeley, 1986.

Beckmann, Petr. *A History of Pi,* Boulder, Colorado, 1982.

Bowra, Cecil Maurice. "Homer," *The Oxford Classical Dictionary,* Second Edition, Oxford (1970): 524-526.

Boyd, Thomas D. and Michael H. Jameson. "Urban and Rural Land Division in Ancient Greece," *Hesperia* 50 (1981): 327-342.

Broneer, Oscar. *Corinth,* Results of Excavations Conducted by The American School of Classical Studies at Athens, Volume I, Part iv, *The South Stoa and Its Roman Successors,* Princeton, 1954.

_____. *Isthmia,* Excavations by the University of Chicago under the auspices of the American School of Classical Studies at Athens, Volume II, *Topography and Architecture,* Princeton, 1973.

_____. "Excavations at Isthmia, Fourth Campaign, 1957-1958," *Hesperia* 28 (1959): 298-343.

Camp, John M. *The Athenian Agora, Excavations in the Heart of Classical Athens,* New York, 1986.

Catling, Hector. "Archaeology in Greece," *Archaeological Reports for 1980-1981,* British School at Athens, London, 1981.

Chace, Arnold Buffam and Henry Parker Manning, *The Rhind Mathematical Papyrus,* Volume II, Oberlin, 1927.

Chantraine, Pierre. *Dictionnaire étymologique de la langue grecque,* Paris, 1968.

Coulton, J. J. *Ancient Greek Architects at Work,* Ithaca, 1977.

Dakaris, S.I. *Archaeological Guide to Dodona,* translated by Elli Kirk Defteriou, Ioannina, 1971.

Decker, Wolfgang. *Annotierte Bibliographie zum Sport im alten Ägypten,* St. Augustin, 1978.

_____. "Die Lauf-Stele des Königs Taharka," *Kölner Beiträge zur Sportwissenschaft* 13, St. Augustin (1984): 7-37.

_____. *Sports and Games of Ancient Egypt,* New Haven, 1992. Translated from German by Allen Guttmann, *Sport und Spiel im Alten Ägypten,* Munich, 1987.

Delorme, Jean. *Gymnasion. Étude sur les monuments consacrés a L'éducation en Grèce,* Bibliothèque des Écoles Françaises d'Athènes et de Rome, fascicle 196, Paris, 1960.

Dilke, O. A. W. *The Roman Land Surveyors, An Introduction to the Agrimensores,* Newton Abbot, 1971.

_____. *Mathematics and Measurement,* London, 1987.

Dinsmoor, William Bell. *The Architecture of Ancient Greece,* reprint of 1950 Third Edition, revised, New York, 1975

Edwards, I.E.S. *The Pyramids of Egypt,* New York, 1988.

Firth, Cecil M., and J.E. Quibell with plans by J.-P. Lauer, *Excavations at Saqqara, The Step Pyramid,* Cairo, 1935.

Frankfort, Henri. *More Sculpture From the Diyala Region,* The University of Chicago, Oriental Institute Publications, Volume LX, Chicago, 1943.

Frazer, J.G. *Pausanias Description of Greece*, translated with a commentary, Volumes I-VI, London, 1913.

Frisk, Hjalmar. *Griechisches etymologisches Wörterbuch,* Vol. 2, Heidelberg, 1970.

Gardiner, E. Norman. *Greek Athletic Sports and Festivals,* London, 1910.

_____. *Athletics of the Ancient World*, Oxford, 1930 (reprinted, Chicago 1980.)

Gasparri, Carlo. "Lo stadio panatenaico," *Annuario della R. Scuola Archeologica di Atene* 52-53, New Series, 36-37 (1974-1975): 313-392.

Gauer, Werner. "Die Tongefässe aus den Brunnen unterm Stadion-Nordwall und im Südost-Gebiet," *Olympische Forschungen* VIII, Emil Kunze and Alfred Mallwitz, eds., Berlin, 1975.

Gebhard, Elizabeth R. "The Form of the Orchestra in the Early Greek Theater," *Hesperia* 43 (1974): 428-440.

_____. "The Sanctuary of Poseidon on The Isthmus of Corinth and The Isthmian Games," in *Mind and Body, Athletic Contests in Ancient Greece,* Olga Tzachou-Alexandri, ed. (1988): 82-88.

_____. "The Early Stadium at Isthmia and the Founding of the Isthmian Games," *Proceedings of an International Symposium on The Olympic Games* (1988), William Coulson and Helmut Kyrieleis, eds, Athens (1992): 73-79.

_____. and Frederick P. Hemans. "University of Chicago Excavations at Isthmia, 1989: I, *Hesperia* 61 (1992): 1-77.

Goodyear, William Henry. *Greek Refinements: Studies in Tempermental Architecture,* New Haven, 1912.

Graef, P. in Curtius and Adler, eds., "Das Gymnasion," *Olympia, Die Ergebnisse der von dem deutschen Reich veranstalteten Ausgrabung,* Vol. II: *Die Baudenkmaler von Olympia,* Berlin 1892.

Hampe, Roland and Ulf Jantzen. "Verschiedene Weihgaben," *OlBer* I, Berlin (1937): 77-82.

Harris, H.A. "Stadia and Starting Grooves," *Greece and Rome,* Second Series (1960): 25-35.

_____. *Greek Athletes and Athletics,* London, 1964.

Haselberger, Lothar and Hans Seybold. "Seilkurve oder Ellipse? Zur Herstellung antiker Kurvaturen nach dem Zeugnis der Didymeischen Kurvenkonstruktion," *Archäologischer Anzeiger* (1991): 165-188.

Heath, Sir Thomas. *A History of Mathematics,* Volume 1, *From Thales to Euclid,* Oxford, 1921 (republished New York 1981.)

Heilmeyer, Wolf-Dieter. "Durchgang, Krypte, Denkmal: zur Geschichte des Stadioneingangs in Olympia," *AM* 99 (1984): 251-263.

Herbert, Sharon Carey. "Corinthian Red Figure Pottery," a dissertation in Classics, Stanford University, 1972.

Herrmann, Klaus. "Olympia, The Sanctuary and the Contests," *Mind and Body, Athletic Contests in Ancient Greece,* Olga Tzachou-Alexandri (1988): 47-68.

Hönle, Augusta and Anton Henze. *Römische Amphitheater und Stadien, Gladiatorenkämpfe und Circusspiele,* Zurich, 1981.

Hultsch, Friedrich. *Griechische und römische Metrologie,* Berlin, 1882.

Humphrey, John. *Roman Circuses, Arenas for Chariot Racing,* Berkeley, 1986.

Hyde, Walter Woodburn. *Olympic Victor Monuments and Greek Athletic Art,* Washington, D.C., 1921.

Jameson, Michael H. "Excavations at Porto Cheli, Excavations at Halieis, Final Report," *Deltion* 27, 1972, Chronika, pp. 233-236.

_____. "Halieis at Porto Cheli," in D.J. Blackman, ed., *Marine Archaeology,* Hamden, Conn. (1973): 219-229.

_____. "The Excavation of a Drowned Greek Temple," *Scientific American,* October (1974): 111-119.

_____. "The Submerged Sanctuary of Apollo at Halieis in the Argolid of Greece," *National Geographic Society Research Reports,* volume 14, Washington, D.C. (1982): 363-367.

Jannoray, Jean. "Le Gymnase," *Fouilles de Delphes,* Vol. II, Topographie et Architecture, Part III, Paris, 1953.

Jüthner, Julius. "Dromos 2," *Pauly-Wissowa, Real-Encyclopädie der klassischen Altertumswissenschaft,* V 2, Stuttgart (1905): cols. 1717-1720.

_____. "Stadion (lauf)," *Pauly-Wissowa, Real-Encyclopädie der klassischen Altertumswissenschaft,* III A 2, Stuttgart (1929): cols. 1963-1966.

_____. and Friedrich Brein, ed. *Die Athletischen Leibesübungen der Griechen,* Vols. I, II. Vienna, 1965, 1968.

Kavvadias, P. "Τὸ Στάδιον, Ἀνασκαφαὶ ἐν Επιδαύρωι, *Praktika* (1900-1902): 78-92.

Kemp, Barry J. "A Building of Amenophis III at Kôm El-'Abd," *Journal of Egyptian Archaeology* 63 (1977): 71-82.

_____. *Ancient Egypt, Anatomy of a Civilization,* New York, 1989.

Kourouniotes, K. "Ἀνασκαφαὶ Λυκαίου," *Praktika* (1909): 185-200.

Knackfuss, Hubert. Milet Vol. 1.7, *Der Sudmarkt und die benachbarten Bauanlagen,* Berlin, 1924.

Knorr, Richard Wilbur. *The Ancient Tradition of Geometric Problems,* Boston, 1986.

Koenigs, Wolf. "Stadion III und Echohalle," in *Olympia Berichte* X, Deutsches Archäologisches Institut, Berlin (1981): 353-369.

_____. *Die Echohalle,* Deutsches Archäologisches Institut, Berlin, *Olympische Forschungen* 14, 1984.

Kramer, Samuel Noah. *History Begins at Sumer,* Third revised edition, Philadelphia, 1981.

Kunze, Emil. "Das Stadion," *Olympia Berichte* V, Deutsches Archäologisches Institut, Berlin (1956): 10-34.

_____. and Hans Schleif, "Das Stadion," *Olympia Berichte* II, Deutsches Archäologisches Institut, Berlin (1938): 8-12.

_____. and Hans Schleif, "Das Stadion," *Olympia Berichte* III, Deutsches Archäologisches Institut, Berlin (1939): 10-12.

_____. and Hans Schleif. "Das Gymnasion," *OlBer* III (1941): 67-75.

Kyle, Donald G. *Athletics in Ancient Athens,* Leiden, 1987.

Lauer, Jean-Philippe, *Fouilles à Saqqarah, La Pyramìde à Degrés, L'Architecture,* Paris 1936.

Lee, Hugh M. "The 'First' Olympic Games of 776 B.C.," in Wendy J. Raschke, ed., *The Archaeology of The Olympics,* Madison, Wisconsin (1988): 110-118.

Leon, Veronica Mitsopoulos- and Erwin Pochmarski. "Elfter Vorlaufiger Bericht über die Gräbungen in Elis," *ÖJh* 51 (1976-1977): cols. 181-222.

Lehman-Haupt, . "Stadion (etymologie, metrologie), "*Pauly-Wissowa, Real-Encyclopädie der klassischen Altertum-wissenschaft,* III A2, Stuttgart (1929), cols. 1930-1963.

Mallwitz, Alfred. "Das Stadion," *Olympia Berichte* VIII (1967): 16-82.

_____. *Olympia und Seine Bauten,* Munich, 1972.

_____. "Zu den Arbeiten im Heiligtum von Olympia während der Jahre 1967-1971," *ArchDelt* 27 (1972), *Chronika,* pp. 273-276, plates 211-212.

_____. "Cult and Competition Locations at Olympia," in *The Archaeology of The Olympics,* ed. by Wendy J. Raschke, Madison, Wisconsin (1988): 79-109.

_____ and Hans-Volkmar Herrmann, eds., *Die Funde aus Olympia,* Athens, 1980.

Miller, Stella G. "Excavations at Nemea, 1982," *Hesperia* 52 (1983): 70-95.

Miller, Stephen G. "Excavations at Nemea, 1977," *Hesperia* 47 (1978): 1-26.

_____. "Turns and Lanes in the Ancient Stadium," *American Journal of Archaeology* 84 (1980): 159-166.

_____. ed. *Nemea, A Guide to The Site and Museum,* Berkeley, 1990.

_____. "The Stadium at Nemea and the Nemean Games," *Proceedings of an International Symposium on the Olympic Games (1988),* William Coulson and Helmut Kyrieleis, eds., Athens (1992): 81-86.

Moretti, Luigi. *Olympionikai. I vincitori negli antichi agoni Olimpici,* Rome, 1957.

_____. "Supplemento al Catalago degli Olympionikai, "*Klio* 52 (1970): 295-339.

_____. "Nuovo supplemento al catalogo degli Olympionikai," *Proceedings of an International Symposium on the Olympic Games (1988),* William Coulson and Helmut Kyrieleis, eds., pp. 119-128.

Morgan, Catherine. *Athletes and Oracles, The Transformation of Olympia and Delphi in the Eighth Century B.C.,* Cambridge, 1990.

Morgan, II, Charles. "Excavations at Corinth, 1936-1937," *American Journal of Archaeology* 41 (1937): 539-552.

Moussa, Ahmed M. "A Stele of Taharqa from the Desert Road at Dahshur," *Mitteilungen des Deutschen Archäologischen Instituts, Abteilung Kairo,* 37 (1981): 331-337.

Mylonas, Paul. Περὶ Σταδίων, Athens, 1952.

Nisetich, Frank J. *Pindar's Victory Songs,* Baltimore, 1980.

Neudecker, Richard. *Die Skulpturenaustattung römischer Villen in Italien,* Mainz am Rhein, 1988.

Neugebauer, O. *The Exact Sciences in Antiquity,* Second Edition, New York 1969.

O'Connor, David. "Boat Graves and Pyramid Origins, New Discoveries at Abydos, Egypt," *Expedition,* 33, no. 3 (1991): 5-17.

Page, D.L. *Poetae Melici Graeci,* Oxford, 1962.

Patrucco, Roberto. *Lo sport nella Grecia antica,* Florence, 1972.

_____. *Lo stadio di Epidauro,* Florence, 1976.

Pierart, Marcel and Jean-Paul Thalmann. "Agora: zone du Portique," *BCH* 102 (1978): 777-783.

Poliakoff, Michael B. *Combat Sports in the Ancient World: Competition, Violence and Culture,* New Haven, 1987.

Raschke, Wendy J., ed., *The Archaeology of The Olympics,* Madison, Wisconsin, 1988.

Raubitschek, Antony E. *Dedications From The Athenian Akropolis*, Cambridge, Mass. 1949.
_____. "The Agonistic Spirit in Greek Culture," *The Ancient World* 7 (1983): 3-7.
Romano, David Gilman. "An Early Stadium at Nemea," *Hesperia* 46 (1977): 27-31.
_____. "The Stadia of the Peloponnesos," dissertation in Classical Archaeology, University of Pennsylvania, 1981, University Microfilms, Ann Arbor, Michigan.
_____. "The Stadium of Eumenes II at Pergamon," *American Journal of Archaeology* 86 (1982): 586-589.
_____. "The Ancient Stadium: Athletes and Arete," *The Ancient World* 7 (1983): 9-16.
_____. "The Panathenaic Stadium and Theater of Lykourgos: A Re-examination of the Facilities on the Pnyx Hill," *American Journal of Archaeology* 89 (1985): 441-454.
_____., ed. "Exploring 5000 Years of Athletics," *Expedition* 27, no. 3 (1985).
_____. and Benjamin C. Schoenbrun. "A Computerized Architectural and Topographical Survey of Ancient Corinth," *Journal of Field Archaeology* 20 (1993): 177-190.
Roos, Paavo. "The Start of the Greek Foot Race," *Opuscula Atheniensia* 6 (1965): 149-156.
_____. "Labraunda och de antika stadionanlaggningarna," *Svenska Forskningsinstitutet i Istanbul Meddelanden* 2 (1977): 23-29.
_____. "Wiederverwendete Startblöcke vom Stadion in Ephesos," *Jahreshefte des österreichischen archäologischen Instituts,* 52 (1978-): 109-113.
Salmon, J. B. *Wealthy Corinth*, Oxford, 1984.

Sandys, Sir John. *The Odes of Pindar, Including Principal Fragments*, Cambridge, 1968.
Scanlon, Thomas F. "The Footrace of the Heraia at Olympia," *The Ancient World* 9 (1984): 77-90.
_____. *Greek and Roman Athletics, A Bibliography*, Chicago, 1984.
Schilbach, Jürgen. "Olympia, die Entwicklungsphasen des Stadions," *Proceedings of an International Symposium on the Olympic Games (1988),* William Coulson and Helmut Kyrieleis, eds., Athens (1992): 33-37.
Schone, H., ed., *Hero of Alexandria,* Vol. III, *Metrica, Dioptra, Teubner,* 1903.
Shear, T. Leslie Jr. "The Athenian Agora: Excavations of 1973-1974," *Hesperia* 44 (1975): 331-374.
Sjöberg, Åke W. "Trials of Strength, Athletics in Mesopotamia," in "Exploring 5000 Years of Athletics, David Gilman Romano, ed., *Expedition* 27, no. 3 (1985): 7-9.
Sommella, P. "Stadio," *Enciclopedia dell' arte antica, classica e orientale*, Vol. II: 464-468.
Sweeney, Jane, Tam Curry and Yannis Tzedakis, eds., *The Human Figure in Early Greek Art,* Athens, 1987.
Thulin, Carolus, ed., *Corpus Agrimensorum Romanorum,* i I, Teubner, Leipzig, 1913.
Tobin, Jennifer. "The Monuments of Herodes Atticus," dissertation in Classical Archaeology, University of Pennsylvania, 1991.
_____. "Some New Thoughts on Herodes Atticus's Tomb, His Stadium of 143/4, and Philostratus VS 2.550," *American Journal of Archaeology* 97 (1993): 81-89.
Touny, A.D. and Steffen Wenig. *Der Sport in Alten Ägypten,* Lausanne, 1969.

Travlos, John. *Pictorial Dictionary of Athens,* London, 1971.

Tzachou-Alexandri, Olga, ed. *Mind and Body, Athletic Contests in Ancient Greece,* Athens, 1988.

Vanhove, Doris, ed. *Le Sport dans la Grèce Antique, du Jéu à la Compétition,* Brussels, 1992.

Welch, Katherine. "Roman amphitheatres revived," *Journal of Roman Archaeology* 4 (1991): 272-281.

Wiegand, Theodor. *Didyma* I, Berlin, 1941.

_____ and Hans Schrader. *Priene,* Berlin, 1904.

Will, Edouard. *Korinthiaka, Recherches sur l'histoire et la civilisation de Corinthe des Origines aux Guerres Médiques,* Paris, 1955.

Williams, Charles K., II. "Corinth, 1969: Forum Area," *Hesperia* 39 (1970): 1-39.

_____. "Corinth 1978: Forum Southwest," *Hesperia* 48 (1979): 105-144.

_____. "Pre-Roman Cults in the Area of the Forum of Ancient Corinth," dissertation in Classical Archaeology, University of Pennsylvania, 1978.

_____. "Review of J.J. Coulton, *Ancient Greek Architects at Work," The Art Bulletin* 62 (1980): 151-153.

_____ and Joan E. Fisher. "Corinth, 1970: Forum Area," *Hesperia* 40 (1971): 1-51.

_____ and Pamela Russell, "Corinth Excavations of 1980," *Hesperia* 50 (1981): 1-44.

Wojcik, Maria Rita. *La Villa dei Papiri ad Ercolano,* Rome 1986.

Zschietzschmann, Willy. *Wettkampf und Übungstätten in Griechenland,* I, Das Stadion, Stuttgart, 1960.

Index